Lecture Notes of the Institute for Computer Sciences, Social Informatics and Telecommunications Engineering   59

Maarten H. Lamers   Fons J. Verbeek (Eds.)

Human-Robot Personal Relationships

Third International Conference, HRPR 2010
Leiden, The Netherlands, June 23-24, 2010
Revised Selected Papers

 Springer

Volume Editors

Maarten H. Lamers
Fons J. Verbeek

Leiden University
Institute of Advanced Computer Science (LIACS)
Niels Bohrweg 1, 2333CA Leiden, The Netherlands
E-mail: {lamers; fverbeek}@liacs.nl

ISSN 1867-8211   e-ISSN 1867-822X
ISBN 978-3-642-19384-2   e-ISBN 978-3-642-19385-9
DOI 10.1007/978-3-642-19385-9

Springer Heidelberg Dordrecht London New York

Library of Congress Control Number: 2011921626

CR Subject Classification (1998): I.2, H.5, I.4, H.4, C.2, F.1

Typesetting: Camera-ready by author, data conversion by Scientific Publishing Services, Chennai, India

Printed on acid-free paper

Springer is part of Springer Science+Business Media (www.springer.com)

Preface

After two successful editions of the HRPR conference, it was a challenge to meet the high expectations that were raised. This challenge contributed to and fueled the organizational and scientific work that made HRPR 2010, the Third International Conference on Human – Robot Personal Relationships, the success it became.

Since long ago, thoughts of personal relationships between man and artificial beings have been food for myths, speculation, fear, ridicule, entertainment, and science. Advances in technology and science, but also public interest, are making artificial partners increasingly likely in any of the many forms imaginable. Increasingly, researchers from scientific fields such as (social) robotics, human – computer interaction, artificial intelligence, psychology, philosophy, sociology, and theology are involved in their study.

HRRP 2010 aimed at bringing together international researchers, developers, and users to discuss issues and trends, recent research, technological advances, and experiences related to personal relationships with artificial partners. All facets of such relationships were considered relevant – their formation, possibilities, reality, and consequences.

The conference was organized as a single-track, multi-session event. To stimulate interaction and the formation of research collaborations, the atmosphere was deliberately kept informal so as to create an open-minded ambiance. To this effect, the venue was part of the historic academic center of Leiden and the social program started on the evening prior to the conference opening. Particular effort was made to interest and enable students to participate, i.e., by offering them very low registration fees.

Contributions to HRPR 2010 were solicited in the form of original papers (describing original research or design work), position papers (posing substantiated opinions or positions), extended abstracts (describing original and ongoing work), workshop proposals, demonstrations of running system prototypes, and even artistic installations. Submitted contributions were selected by peer-review.

A total of 22 papers and extended abstracts were submitted. Of these, 16 were selected and included in the final program. Combined with one keynote presentation and two invited speakers, eight sessions were scheduled over two days. Among the different session themes were a sociological outlook on the field, design issues, robots and children, robots in care, and artistic approaches. Presentations were followed by lively discussion, from which the breadth of the approaches became apparent but not problematic. Perhaps a good example of this was the positive response to theologist William David Spencers' presentation on *digital adultery*.

Keynote speaker and a pioneer in the research field Kerstin Dautenhahn shared her view on current developments. From both her presentation and the overall conference, it is apparent that defining and structuring of related issues is of priority, and that advances toward the few shared goals are made in small steps only. To kindle reflection on the topic of human – robot relationships from a research-external view, invited artist Edwin van der Heide presented his experiences with an autonomous interactive installation.

With respect to the continuation of the conference series, it is to the community's great pleasure that the Institute for Computer Sciences, Social Informatics and Telecommunications Engineering (ICST) has agreed to sponsor future HRPR editions. From HRPR 2010 we look forward to the *First ICST Conference on Human-Robot Personal Relationships*, HRPR 2011.

Organization

Program Committee

Maarten Lamers (Chair)	Leiden University Institute of Advanced Computer Science, The Netherlands
Fons Verbeek (Vice Chair)	Leiden University, The Netherlands
Ronald Arkin	Georgia Institute of Technology, USA
Joost Broekens	Delft University of Technology, The Netherlands
Jaap van den Herik	Tilburg University, The Netherlands
Bernhard Hommel	Leiden University, The Netherlands
Stefan Kopp	Bielefeld University, Germany
David Levy	Author of "Love and Sex with Robots: The Evolution of Human-Robot Relationships"; winner of the 2009 Loebner Prize in Artificial Intelligence
Cees Midden	Eindhoven University of Technology, The Netherlands
Bernhard Sendhoff	Honda Research Institute Europe
Britta Wrede	Bielefeld University, Germany

Table of Contents

Loving Machines: Theorizing Human and Sociable-Technology Interaction

Glenda Shaw-Garlock

School of Communication, Faculty of Communication,
Arts & Technology, Simon Fraser University, 8888 University Drive,
Burnaby, BC, Canada
grshawga@sfu.ca

Abstract. Today, human and sociable-technology interaction is a contested site of inquiry. Some regard social robots as an innovative medium of communication that offer new avenues for expression, communication, and interaction. Other others question the moral veracity of human-robot relationships, suggesting that such associations risk psychological impoverishment. What seems clear is that the emergence of social robots in everyday life will alter the nature of social interaction, bringing with it a need for new theories to understand the shifting terrain between humans and machines. This work provides a historical context for human and sociable robot interaction. Current research related to human-sociable-technology interaction is considered in relation to arguments that confront a humanist view that confine 'technological things' to the nonhuman side of the human/nonhuman binary relation. Finally, it recommends a theoretical approach for the study of human and sociable-technology interaction that accommodates increasingly personal relations between human and nonhuman technologies.

Keywords: automatons, social robots, human robot interaction, actor-network theory.

1 Introduction

This work seeks to accomplish two exploratory objectives. The first objective involves situating emerging human and sociable-technology interaction as a developing site of inquiry relevant to communication studies and cultural studies. To this end it (briefly) overviews the historical and contemporary context of social robots. The second objective involves theorizing increasingly affective human and sociable-technology interaction by drawing on approaches that accommodate increasingly intimate relations between human and nonhuman (sociable) technologies. To this end, current research which interrogates "sociableness" between human and sociable-technology [4, 25, 26, 43, 44] is considered in relation to arguments that confront a humanist view that confine 'technological things' to the nonhuman side of the human/nonhuman binary relation [16, 28]. As intimate machines become increasingly capable of engaging people in affective social relationships, unfamiliar moments emerge as humans and nonhumans comingle in ways that make Western moderns feeling uncomfortable and call for improved frameworks for coming to terms with the shifting relationship between people

M.H. Lamers and F.J. Verbeek (Eds.): HRPR 2010, LNICST 59, pp. 1–10, 2011.
© Institute for Computer Sciences, Social Informatics and Telecommunications Engineering 2011

and machines. Today, human and sociable-technology interaction is a contested site of inquiry, with some regarding social robots as an exciting medium of communication that will ultimate shift the way we view ourselves and others. Still, others question the moral veracity of human-robot relationships, suggesting that such associations risk psychological impoverishment [19] or are disconcertingly inauthentic [34, 39] and therefore morally problematic [43]. What seems clear is that the emergence of social robots in everyday life will alter the nature and dynamics of social interaction [48], bringing with it a need for new theories to understand the shifting terrain between humans and machines.

2   Historical Context of Social Technologies

Today, many people are familiar with the technological figure of the social robot which is purposefully designed to engage human users in emotional relationships. Indeed, the quest to (re)create a perfect version of "man" has engaged the imagination of philosophers, inventors, and scientists since Antiquity [7, 8].[1] Less acknowledged however, is the continuity that exists between eighteenth-century automatons and twenty-first century humanoid social robots [35]. For example, from the beginning interaction between humans and automatons was inflected with intense, sometimes conflicting, emotions: from triumphant feelings of technological achievement [14], to erotic passion [33], to disorienting feelings of intellectual uncertainty [17].

This paper regards the mid eighteenth-century as a defining moment in the history of the quest for artificial life. During this period, automata were considered to be the very symbol of the Enlightenment, with Jacques de Vaucanson widely regarded as the "forerunner of the great eighteenth-century builders of automata" ([7], p. 275). The eighteenth-century also represents a critical moment wherein, "...the ambitions of the necromancers were revived in the well-respected name of science...an interest in anatomy, advances in the design of scientific instruments...meant that automata were thought of as glorious feats of engineering, or philosophical toys" ([47], p. xvi).

During this period, mechanicians produced life sized highly sophisticated humanoid automata that played music using fingers, lips and breath; drew detailed portraits of royal subjects; and scribed such intellectual poetry as, "I do not think, therefore am I not?" The human actions mechanicians (automaton builders) selected to simulate with their automata were, at the time, regarded as the very height of human essence (e.g. music, poetry, art). The automatons of Vaucanson and others were presented at courts, fairs, and exhibitions throughout Europe. Parisans crowds were variously delighted and unsettled by these ingenious mechanisms. Yet, in

---

[1] The first mechanical automata for which there are verifiable records emerged around second or third century BC with the invention of mechanical water-clocks of Ctesibius, Philo the Byzantine, and Hero of Alexandria who collectively represented the Alexandrian School. The sciences of the Alexandrians were preserved through translations authored by Hero of Alexandria into Arabic and Latin at Byzantium and from the sixteenth century onward, authors remained inspired by the science of the Alexandrians, but added personal ideas and inventions to these early inventions applying the art and science of hydraulic automata to the gardens and grottoes of princes and kings during the sixteenth and seventeenth century ([5], p. 31-36).

addition to being wildly popular entertainment artifacts, these automata were boldly engaged with the central debates of their time. "Vaucanson's automata were philosophical experiments, attempts to discern which aspects of living creatures could be reproduced in machinery, and to what degree, and what such reproductions might reveal about their natural subjects" ([31], p. 601). So we can see that the blurring of the boundary between human and machine; the evocation of feelings associated with the uncanny; the ambivalence felt toward the steady march of scientific progress; and the engagement with the central debates of their time are attributes that the eighteenth-century automaton and contemporary artificial life projects seem to share.

With rise of industrial development (1760 onward), Western Europe and North America were transformed and the age of the automata drew to a close and with it the quest to simulate life was abandoned. In 1847, Helmholtz, reflecting on Vaucanson's automata, mused, "nowadays we no longer attempt to construct beings able to perform a thousand human actions, but rather machines able to execute a single action which will replace that of thousands of humans" ([2], p. 41). In the historical record life-like automata are regarded as the "progenitors of the Industrial Revolution" [2, 10]; they were the concrete manifestation of a nation's scientific ability, "[embodying] what was, at the time, the absolute cutting edge of new technology" ([38], p. 2) and provided the technological foundation for the advancing Industrial Revolution. Amusement automata, "capable of self-government and intelligence" ([33], p. 335) assisted the vision of this new factory system.

If the eighteenth century was driven by the desire to know if life could be reproduced mechanically, our contemporary moment is characterized by a resolve that we can and will create intelligent machines (shifting from the idea of human-as-machine to machine-as-human) that will be endowed with artificial intelligence, consciousness, free will, and autonomy [45].

As a site of inquiry relevant to communication and cultural studies, a handful of media scholars are beginning to consider the social robot as a unique medium of communication that may ultimately affect the way that we see ourselves and relate to others and "extend new possibilities for expression, communication and interaction in everyday life" ([29], p. 328, [48]). Within cultural studies, Thrift [39] suggests that social robots represent a site beyond conventional structures and sites of communication and is concerned with issues of authenticity, while Roderick [32] examines representations of agency and automation of Explosive Ordnance Disposal (EOD) robots in mass media. Finally, I note that human-robot interaction (HRI) studies [24] concerned with the practical and theoretical implications for the design of social robots are finding their way into top tier communication journals.

3 Social Robotics: Sociable, Relational and Present

Variously referred to as friendly machines, socially intelligent robots [26], relational artifacts [40, 41, 43], and robotic others [19], the social robot is designed to give the impression "of wanting to be attended to, of wanting to have their 'needs' satisfied, and of being gratified when they are appropriately nurtured" ([43], p. 331). Social robots are technologies designed to engage with humans on an emotional level through play, sometimes therapeutic play, and perhaps even companionship.

Examples of commercially available entertainment robots include, Innvo Lab's robotic dinosaur *Pleo*, Tiger Electronic's hamster-like *Furby* and Sony's puppy-like robot *AIBO*. Japan's National Institute of Advanced Industrial Science and Technology created *Paro*, a harp seal robot, to serve as companions for Japan's growing senior citizen population [18] and as therapeutic playmates for children with autism [9]. Highly advanced social robots include MIT's *Kismet*, *Cog* and *Nexi* as well as Osaka University's *Repliee Q1* and *Repliee Q2*.

According to social robot creator, Cynthia Breazeal, "...a sociable robot is able to communicate and interact with us, understand and even relate to us, in a personal way" ([4], p. 1). Fong, Nourbakhsh, and Dautenhahn [13] envision social robots as "embodied agents" belonging to a heterogeneous group. They would recognize and engage one another in social interaction and possess histories and "perceive and interpret the world in term of their own experience" (p. 144).

Sherry Turkle refers to this category of robot as relational artifacts, defined as "artifacts that present themselves as having 'states of mind' for which an understanding of those states enriches human encounters with them" ([43], p. 347). The term 'relational' also denotes her psychoanalytic orientation and an emphasis upon human-meaning within user-technology interaction.

Hiroshi Ishiguro is a pioneer in researching the significance of the aesthetic appearance of social robots upon human-robot interactions. Ishiguro is interested in the conveyance of *sonzai-kan*, or human presence, and the best way to evoke the sensation of "presence" within the social robot's human social partner. "Simply put, what gives something a social presence? Is it mainly behavior, or is there instead some complex interplay between appearance and behavior?" ([27], p. 1) For Ishiguro appearance *and* behavior are the critical factors in creating sufficiently believable humanoid robots.

The idea that a nonhuman figure, a robot, might possess presence or engage human interactants in emotional relationships highlights the receding boundary between humans and machines. Donna Haraway [16] argues that people in Western society have gradually become biotechnological hybrids, and as such are all cyborg like beings, a fact that destabilizes dualisms underpinning much of Western thinking. Consequently, dichotomies which bind and separate the world, like the mind/body split, human/machine, Self/Other, male/female are rendered meaningless and irrelevant.

Propelled by the radical ideas of Descartes,' eighteenth century mechanicians (automata builders) were early problemitizers of Western dualism, as they sought to simulate as closely as possible both the external and internal mechanisms of life. Jessica Riskin [31] argues that the *age of the automaton* differs from periods immediately preceding and following it in that this period was concerned with collapsing the perceived differences between humans and machines rather than maintaining distance.

4 Tensions: Authenticity and New Technological Genres

"Our culture [has] clearly come to a new place" muses Sherry Turkle after reflecting on yet another threshold crossed in ongoing cultural-technical imaginary suggested in the release of David Levy's *Love and Sex With Robots* [25]. As the relationship between humans and social robotics become increasing diverse and complex uncomfortable questions begin to emerge relating to what it means to engage in authentic social relationships [11, 43], what it means to love [25], what it means to be human (and machine) [20] as well as all the attendant ethical concerns [1, 34, 36, 37] related to these apprehensions. Anxiety about authenticity and deception as well as the emergence of novel ontological categories of 'being' are briefly outlined below and where ever possible connected to similar concerns of the eighteenth century period of mechanistic simulation.

Authenticity & Deception. A study by Turkle [44] examined the interaction between senior citizens and Paro (a robotic baby harp seal) and found that Paro successfully elicited feelings of admiration, loving behavior, and curiosity but felt that these interactions raised thorny "questions about what kind of authenticity we require of our technology. Do we want robots saying things that they could not possibly 'mean'? What kinds of relationships do we think are most appropriate for our children and our elders to have with relational artifacts?" ([44], p. 360). Sharkey and Sharkey [34] raise parallel concerns when they consider the natural human tendency to anthropomorphize and suggest we question the ethical appropriateness of deceiving people into believing that a machine is capable of mental states and emotional understanding. Kahn et al. [19] question a sociable robot's ontological status as 'social' and their ability to engage in truly social behavior, doubting that intelligent machines can ever really interpret the world around them in terms of their own experience.

On the other hand, Duffy [12] points out that from the point of view of social robotics, it doesn't really matter whether or not a particular robot *genuinely* possesses a sense of personal agency, intention, or self awareness. What matters most is our perception of their emotionality and intelligence. Further, as Nass and Moon's [30] research has shown, in spite of the fact that most technology-users consciously view machines as non-persons, they persistently engage in behavior that may best be described as ethopeoia, involving "a *direct* response to an entity as human while knowing that the entity not warrant human treatment or attribution" ([30], p. 94). Turkle suggests that relational artifacts are so successful at engaging humans in emotional relationships because they successfully press our Darwinian buttons through eye contact, gesture, vocalization, and so on [42].

Contemporary thinkers' concerns about authenticity and deceit in relation to social robots resonate with eighteenth century concerns over similar mechanistic simulations of life. For example, when eighteenth century automata builder Jacque de Vaucanson first revealed his *Flute Player* and *Tabor-and-Tambourine Player* to the Academie des Sciences in Paris in 1738, spectators were profoundly suspicious of his mechanisms, "At first many people would not believe that the sounds were produced by the flute which the automaton was holding" ([7], p. 274). However, after thorough inspection Voltaire declared the inventor: "bold Vaucanson, rival of Prometheus!" ([38], p. 11) Vaucanson's work provides a dramatic representation of a philosophical preoccupation engaging laymen, philosophers, and royalty throughout this period: the problem of whether or not human processes and functions were essentially mechanical.

Technological Beings & New Ontological Categories. Kahn et al. [20] posit nine
characteristics that social robot designers might strive towards to achieve successful
social acceptance. Similarly, Duffy [10] suggests eight criteria for successful human-
social robot interaction. Yet even without fully attaining all of these dimensions, the
affect of minimal social cues from consumer robots (e.g. Tamagotchi, Furbies, My
Real Baby, AIBO, Pleo) have shifted our acceptance and perception of social robots
dramatically over the past decade and a half.

Sherry Turkle [43, 44] speaks of an emerging category of "being" expressed by
young children engaging with very rudimentary and very advanced social robots,
discovering a tendency to ascribe social robots to an emergent category of *being*,
referred to as "sort of alive," situated between alive and not alive (Kahn et al. [14] call
this a 'new technological genre').[2] Further, Turkle notes that the child of today still feels
compelled to classify social robots (as they did twenty years ago) when first
encountering intelligent machines. However, today this urge to classify is now
increasingly entangled with a desire to nurture and be nurtured by relational
artifacts [43].

Even designers form enduring emotional connections to their projects. Breazeal
admits to being quite attached and never tiring of her interaction with, Kismet, an
engaging social robot. "To me Kismet is very special because when you interact with
Kismet, you feel like you're interacting with Kismet. You know, there's someone
home so to speak [laughs]. There's someone behind those eyes that you're interacting
with." This capacity of social robots to provide a good 'fake' [12] or 'cheat' [11]
leads some critics to conclude that the use of social robots with our most vulnerable
(children and elderly) is potentially unethical, "as it is akin to deception" ([37],
p. 148).

That Kuhn and colleagues [19, 20] favor the term robotic *other* foregrounds their
view that there is a need for new language and words that accurately capture the
shifting ontological status of machines and the new social relationships arising from
our increasing interaction with robotic others. For Turkle, and Donna Haraway as
well [16], this intermediate category is illustrative of a culture in which the boundary
between the inanimate and animate has significantly eroded. It also signals a moment
in which children, adults, and seniors comfortably engage emotionally and
intellectually with increasingly social technologies.

Ronald Arkin states, that "Robotics researchers make a tacit assumption that the
creation of this new technology is wholly appropriate and can only enrich the lives of
those on the receiving end. It is important that we as scientists re-examine this
assumption" ([1], p. 3). What approaches might assist roboticists, engineers, and
computer scientists respond to such complex questions and concerns as the
appropriateness of human and robotic love? How do we begin to assess authenticity
and fully understand and manage anthropomorphism in social robots? What seems
clear is that in the context of robo-nannies, therapeutic seals, military bots, and
domestic and surgical robotics there is need for methodological and theoretical
perspectives that enable us to think through techno-cultural hybrid configurations of
people and machines. In short, what is needed is a "theory of machines" [23].

---

[2] The tendency to anthropomorphize and attribute life like essences to social robots is not
limited to children [14, 23].

5 Actor Network Theory: 'A Theory of Machines'

I suggest that a theory from the field of science and technology studies, actor-network theory, may prove to be a useful theoretical orientation for research related to human-social robotic interaction. In particular because many of its central tenets accommodate and presume non human artifacts, especially the machine, are a constitutive aspect of the social world. Actor-network theory is derived from the sociology science and was pioneered by Michael Callon [5] and Bruno Latour [21]. Later works focused on technology and is therefore sometimes regarded as a branch of the social construction of technology school of thought [3].

Actor-network theory rejects the assumption that society is constructed through human action and meaning alone. Rather, it regards social life as being performed or created by actors, some human, some non human, all of which may be 'enrolled' in the creation of *knowledge* that always takes some material form (e.g. patents, scientific papers, agents, social institutions, machines, technologies, and organizations) [6]. This approach argues that human agents as well as machines are all *effects* of networks of diverse (not simply human) materials. In short, the social is not viewed as a strictly human domain but rather, a "patterned network of heterogeneous material" ([23], p. 381) which includes people, machines, text, institutions and more.

In this regard, actor-network theory is radical because "it treads on a set of ethical, epistemological and ontological toes" ([23], p. 383). Specifically, it does not categorize and privilege humans on one hand, and cordon off non human objects on the other. Indeed, actor-network theory contends that the social and the technical are inseparable. Actor-network theory sets out to study the motivations and actions of actors (human and non human artifacts) "who form elements, linked by associations, of heterogeneous networks of aligned interests" ([46], p. 468). In this way actor-network theory is concerned with the mechanics of power.

A major focus of actor-network theory is to strive to reveal and describe the development of stable networks of allied interests and how these networks come to be created and maintained, and conversely, to study those instances where a network has failed to become established. Micro level moments of creation, maintenance, and failure reveal instances of controversy and uncertainty and expose the occasions where "science and technology [is] in the making" [21]. Thus, Latour proposes investigating at least five sites of major uncertainties including: the contradictory identity of actors within an institutional groups (nature of groups); the competing goals for each course of action (nature of action); the open and varied nature of agencies involved (nature of objects); the ongoing dispute between society and the natural sciences (nature of facts); and the types of scientific studies undertaken (nature of the social study of technology) [22].

The term actor-network theory references both a theoretical and methodological approach. It provides the concepts (e.g. actor, actor-network, enrolment, translation, delegates, irreversibility, black box, etc.) through which to view the socio-technical world as well as the elements which need to be revealed in empirical work. Thus, the researcher is encouraged to document network elements including the human and the non human, processes of translation and inscription, the creation of black boxes or immutable mobiles, and the degree of stability and irreversibility of networks and their elements [46].

To follow the theoretical orientation set up by Latour and Callon we must consider concurrently all of the members, organizations, concepts, places and objects that contribute to a particular scientific outcome. What actor-network theorists propose is that we consider all of these dimensions simultaneously as well as their internal relatedness in order to understand how science operates. "We cannot, for example talk about how Pasteur discovers the cause of anthrax without aligning his biological laboratory in Paris, the anthrax bacillus, the rural farms, the cattle, Pasteur, his ideological framework, the hygienists who supported his work, etc." ([15], p. 43)

Similarly, if we want to apply an actor-network approach to social robotics we might find our entry point through the identification of a problem or controversy with which the field is presently grappling ([21], p. 4), perhaps social presence in robots or conceptions of authenticity. From here we turn our attention to the micro-level work of scientists, engineers, and computer scientists. But also in our analytic frame are all of the other members of the heterogeneous network [23] that make up the social robotics field: robots, humans interactants, ideological assumptions, emerging robot-ethics, codes of conduct, representation (design) choices, corporate entrepreneurs, state policy, scientific publications and more.

In closing, I return to a point that I raised earlier in this essay. That is to say, the current push to fashion life-like creatures through science and technology is by no means a new development. Indeed, this project is at least 250 years old and has at least two distinct periods in which the conceptual boundary between human and machine was sufficiently malleable to be subject to negotiation and change. If we want to understand what happened at the start of the industrial revolution that ultimately coincided with the (re)solidification of the conceptual boundary between man and machine as well as the abandoning of the pursuit to simulate life through mechanism we might consider broadening our analytic frame to include an investigation of all the "bit and pieces" from the socio-technical world that make up our heterogeneous technological products.

References

1. Arkin, R.: On the Ethical Quandries of a Practicing. A First Hand Look, Roboticist (2008)
2. Bedini, S.: The Role of Automata in the History of Technology. Technology and Culture 5(1), 24–42 (1964)
3. Bijker, W., Hughes, T., Pinch, T.: The Social Construction of Technological Systems. MIT Press, Cambridge (1987)
4. Breazeal, C.: Designing Sociable Robots. In: Intelligent Robots and Autonomous Agents, MIT Press, Cambridge (2002)
5. Callon, M.: Some Elements of a Sociology of Translation: Domestication of the Scallops and the Fishermen. In: Law, J. (ed.) Power, Action and Belief: A New Sociology of Knowledge, pp. 196–223. Routledge & Kegan Paul, London (1986)
6. Callon, M., Latour, B.: Unscrewing the Big Leviathan, or How Do Actors Macrostructure Reality and How Sociologists Help Them to Do So. In: Knorr-Cetina, K., Cicourel, A. (eds.) Advances in Social Theory: Toward an Integration of Micro and Macro Sociologies, pp. 277–303. Routledge, London (1981)
7. Chapuis, A., Droz, E.: Automata: A Historical and Technological Study. Neuchatel, Switzerland (1958)

8. Cohen, J.: Human Robots in Myth and Science. AS Barnes and Company, South Brunswick (1967)
9. Dautenhahn, K., Billard, A.: Proceedings of the 1st Cambridge Workshop on Universal Access and Assistive Technology (2002)
10. de Solla Price, D.: Automata and the Origins of Mechanism and Mechanistic Philosophy. Tech. & Cult. 5(1), 9–23 (1964)
11. Duffy, B.R.: Anthropomorphism and the Social Robot. Robot & Auton Sys. 42, 177–190 (2003)
12. Duffy, B.R.: Fundamental Issues in Social Robotics. Inter. Rev. of Info. Ethics 6, 31–36 (2006)
13. Fong, T., Nourbakhsh, I., Dautenhahn, K.: A Survey of Socially Interactive Robots. Robot & Auton Sys. 42, 143–166 (2003)
14. Foucault, M.: Discipline and Punish: Birth of the Prison. Random House, New York (1995)
15. Geraci, R.M.: Apocalyptic Ai: Visions of Heaven in Robotics. In: Artificial Intellegence and Virtual Reality, Oxford University Press, Oxford (2010)
16. Haraway, D.: Simians, Cyborgs, and Women: The Reinvention of Nature. Free Association Books, London (1991)
17. Jentsch, E.: On the Psychology of the Uncanny. Angelaki: A New Journal in Philosophy, Literature, and the Social Sciences 2(1), 7–17 (1996)
18. Johnstone, B.: Japan's Friendly Robots. Tech. Rev. 102(3), 64–69 (1999)
19. Kahn, P., et al.: Social and Moral Relationships with Robotic Others? In: Proceedings of the 13th International Workshop on Robot and Human Interactive Communication, September 20-22, pp. 545–550. IEEE Press, Piscataway (2004)
20. Kahn, P., et al.: What Is a Human? Toward Psychological Benchmarks in the Field of Human-Robot Interaction. Interact. Stud. 8(3), 363–390 (2007)
21. Latour, B.: Science in Action. Harvard University Press, Cambridge (1987)
22. Latour, B.: Reassembling the Social: An Introduction to Actor-Network-Theory. Oxford University Press, Oxford (2005)
23. Law, J.: Notes on the Theory of the Actor Network: Ordering, Strategy and Heterogeneity. Systems Practice 5(4), 379–393 (1992)
24. Lee, K.M., Peng, W., Jin, S.-A., Yan, C.: Can Robots Manifest Personality?: An Empirical Test of Personality Recognition, Social Responses, and Social Presence in Human-Robot Interaction. Journal of Communication 56, 754–772 (2006)
25. Levy, D.: Love + Sex with Robots: The Evolution of Human-Robot Relationships. Harper Collins, New York (2007)
26. MacDorman, K., Ishiguro, H.: The Uncanny Advantage of Using Androids in Cognitive and Social Science Research. Interact. Stud. 7(3), 297–337 (2006)
27. MacDorman, K., et al.: Assessing Human Likeness by Eye Contact in an Android Testbed. In: Proceedings of the 27th Annual Meeting of the Cognitive Science Society (2005)
28. MacKenzie, D.A., Wajcman, J.: The Social Shaping of Technology, 2nd edn. Open University Press, Buckingham (1999)
29. Mayer, P.: Computer Media Studies: An Emergent Field. In: Mayer, P. (ed.) Computer Media and Communication: A Reader, pp. 329–336. University Press, Oxford (1999)
30. Nass, C., Moon, Y.: Machines and Mindlessness: Social Responses to Computers. J. of Soc. Iss. 56(1), 81–103 (2000)
31. Riskin, J.: The Defecating Duck, or, the Ambiguous Origins of Artificial Life. Crit. Inqu. 29, 599–633 (2003b)

32. Roderick, I.: Considering the Fetish Value of Eod Robots. Int. J. Cult. Stud. 13(3), 235–253 (2010)
33. Schaffer, S.: Babbage's Dancer and the Impresarios of Mechanism. In: Spufford, F., Uglow, J. (eds.) Cultural Babbage: Technology, Time, and Invention, pp. 52–80. Faber and Faber, London (1996)
34. Sharkey, N., Sharkey, A.: The Crying Shame of Robot Nannies: An Ethical Approach. Interact. Stud. 11(2), 161–190 (2010)
35. Shaw-Garlock, G.: History of the Mechanical Woman: Automaton to Android (2010) (unpublished Paper)
36. Sparrow, R.: The of the Robot Dogs. Ethics & Info. Tech. 4(4), 141–161 (2002)
37. Sparrow, R., Sparrow, L.: In the Hands of Machines? The Future of Aged Care. Mind & Mach. 16, 141–161 (2006)
38. Standage, T.: The Turk: The Life and Times of the Famous Eighteenth Century Chess Playing Machine. Berkley Books, New York (2002)
39. Thrift, N.: Electric Animals. Cultural Studies 18(2), 461–482 (2004)
40. Turkle, S.: Computer as Rorschach. Society 17(2), 15–24 (1980)
41. Turkle, S.: Computer Reticence: Why Women Fear the Intimate Machine. In: Kramarae, C. (ed.) Technology and Women's Voices: Keeping in Touch, pp. 41–61. Routledge & Kegan Paul in association with Methuen, London (1988)
42. Turkle, S.: Simulation Versus Authenticity. In: Brockman, J. (ed.) What Is Your Dangerous Idea? Today's Leading Thinkers on the Unthinkable. Harper Perennial, New York (2007)
43. Turkle, S., Breazeal, C., Daste, O., Scassellati, B.: Encounters with Kismet and Cog: Children Respond to Relational Artifacts. In: Messaris, P., Humphreys, L. (eds.) Digital Media: Transformations in Human Communication, pp. 313–330. Peter Lang Publishing, New York (2006)
44. Turkle, S., Taggart, W., Kidd, C., Daste, O.: Relational Artifacts with Children and Elders: The Complexities of Cybercompanionship. Connect Sci. 18(4), 347–361 (2006)
45. Veruggio, G.: Euron Roboethics Roadmap. In: EURON Roboethics Atelier (2006)
46. Walsham, G.: Actor-Network Theory and Is Research: Current Status and Future Prospects. In: Lee, A., Liebenau, J., DeGross, J. (eds.) Information Systems and Qualitative Research, pp. 466–480. Chapman & Hall, London (1997)
47. Wood, G.: Living Dolls: A Magical History of the Quest for Mechanical Life. Faber and Faber, London (2002)
48. Zhao, S.: Humanoid Social Robots as a Medium of Communication. N. Med. & Soc. 8(3), 401–419 (2006)

Towards a Sociological Understanding of Robots as Companions

Ellen van Oost[1] and Darren Reed[2]

[1] Faculty of Management and Governance, Twente University, The Netherlands
[2] Science and Technology Studies Unit (SATSU), University of York, United Kingdom
e.c.j.vanoost@utwente.nl, djr14@york.ac.uk

Abstract. While Information Communication Technologies (ICTs) have, in the past, primarily mediated or facilitated emotional bonding between humans, contemporary robot technologies are increasingly making the bond between human and robots the core issue. Thinking of robots as companions is not only a development that opens up huge potential for new applications, it also raises social and ethical issues. In this paper we will argue that current conceptions of human-robot companionship are primarily rooted in cognitive psychological traditions and provide important, yet limited understanding of the companion relationship. Elaborating on a sociological perspective on the appropriation of new technology, we will argue for a richer understanding of companionship that takes the situatedness (in location, network and time) of the use-context into account.

Keywords: Social robots; companionship, sociology.

1 Introduction

Much has been made of the future potential of robots. Increasingly, development has turned to what has been termed 'social robotics', wherein robotic devices play social, assistive or therapeutic roles[1,2]. Cynthea Breazeal [3] who developed Kismet one of the earliest social robot in the mid 1990's, defined a sociable robot as one that "is able to communicate and interact with us, understand and even relate to us, in a personal way" (p.1) She sees the pinnacle of achievement robots that "could befriend us, as we could them".

Such attitudes as those expressed by Breazeal, are instructive because they frame up an idealized relationship based upon communicative action between two essentially isolated individuals. This notion points to the heart of a dilemma in that it would seem to suggest a singular relationship between an isolated robotic artifact and an emotional human being. The robot is seen to evoke in the human a vast array of emotions that result in intimate ties between person and robot. These 'single point' framings are a product of a particular disciplinary background due to a reliance on Human Computer Interaction as a foundation, that itself has roots in cognitive psychology and communications models of human interaction. In this paper we move beyond what might be called a 'single point' notion of interaction between human and

M.H. Lamers and F.J. Verbeek (Eds.): HRPR 2010, LNICST 59, pp. 11–18, 2011.

machine, common to the discipline of Human Robot Interaction (HRI) and Human Computer Interaction (HCI) more generally, to one based upon a sociological understanding of robot companions. We note, with Zhao [15], that until now sociologists have paid surprising little attention to social robots. With this paper we aim to argue the relevance of a sociological understanding of social robots. In doing so we will draw on selected literature in the sociological area of Science and Technology Studies, that moves beyond an ethnographic appreciation of social behaviour and contexts.[1] This approach situates technological artefacts within a broader 'actor-network' and prioritises the relational and transformational nature of the interactions between people and things in particular places [11].

The structure of the paper is as follows. First we will provide a concise impression of how the concept of companion figures in current social robot research by describing some recent and ongoing European projects. Then we will argue the relevance of a broader sociological perspective by highlighting the core social and ethical fears that go with these developments. Lastly we will describe a sociological understanding of robot companions and we will discuss its merits for companions design.

2 Companionship in Current European Research Projects

The notion of "robots as companions" is currently often used to frame current research projects. Just to mention some recent and ongoing core European projects: LIREC (Living with Robots and Interactive Companions) and CompanionAble (both FP 7 projects), COMPANIONS (FP 6 project) and the Austrian project C4U (Companions for Users) that is linked to the FP7 project SERA (Social Engagement with a Rabbitic User Interface).[2] Whereas earlier social robot research tended to emphazise either the functional or the affective dimension of social robots [15], the "companion" approach clearly aims to develop robotic devices that combine these two features. In this section we aim to give an impression on how companionship is conceptualized in these research programmes.

The LIREC project is made up of people from the areas of psychology, ethology, human-computer interaction, human-robot interaction, robotics and graphical characters. As we might expect from such a group, the concerns of the collaboration are wide, but we can discern some key features, such as emotional expression and identification [4,5] along with psychological models of empathy [6]. Enabling devices to remembering previous interaction - and hence learn preferences and the line - and forgetting - so as to protect privacy - are combined to help maintain trust between human and robot [9]. Human-animal relations serve as an important model for robotic companionship in LIREC. This model frames the relationships and points to particulare expectations.

The C4U – Companions for Users – project is one of the few HRI studies that explicitly addresses gender issues. Women constitute a large group of potential users (esp. among the elderly), however, little knowledge is available on the possible gendered character of human robot relationships. This project aims to investigate

---

[1] For early incorporation of STS ideas see the work of Jutta Weber [7].
[2] See for more information the following project websites respectively: lirec.eu, project-sera.eu; www.companions-project.org; www.companionable.net

possible gender-specific requirements for integrating companion technologies into female lifestyles and, in the end, developing guidelines for gender-conscious companion technology development. C4U is related to the FP7 European project SERA in which long-term social engagement with robotic devices is the core challenge. The project studies social engagement with an embodied (Nabaztag), task-oriented (physical exercises) interface over time in real life situations.

The long term bonding with a companion agent is too central in the EU Companions project. Companions, in their perspective, are able of "developing a relationship and "knowing" its owners preferences and wishes". This project focuses on developing meaningful interaction by speech to a central way to establish lasting bonding between humans and virtual companion agents.

CompanionAble, another FP 7 EU project, develops a robotic device that cooperates with Ambient Assistive Living environment, aiming to support longer independent living of (mentally impaired) elderly. The autonomously moving robotic companion mediates between the care-recipient and the smart home environment and care givers. The companion robot is able to detect and track people in the home environment and communicates by speech and (touch) screen. The project uses a co-designing approach by actively involving care-recipients and professional as well as personal care givers during design. Testing prototypes is located in smart home labs mimicking real daily living.

All in all, we may conclude that the concept "companion" is currently featuring in various research projects, yet with multiple meanings. One central issue, however, is that current design aims of artificial companions go beyond "user acceptance" and aim to establish lasting social bonding with its user. As bonding and companionship need time and context to evolve, these projects tend to study the interaction beyond the restricted laboratory setting but in (simulated) real life settings. In this respect we may conclude that current HRI research projects tend to move beyond the traditional "single point" interaction and broaden insights in interaction dynamics by framing the interaction in time and real life contexts. However, most projects still focus primarily on the bilateral human robot interaction leaving the wider social context with different actors and stakeholders involved out of scope. This is especially the case where robot companions are aimed to function in more complex, organizational care arrangements. In the next section we will discuss some of the social and ethical issues that currently rise about the use of robot companions in care settings.

3   Social and Ethical Dimensions of Robots as Companions

The idea of robot companions, while a staple of fiction, is moving into the mainstream. In the application domain for care, there are two primary prospective cohorts for companion robots, children and older people. Not only are there hopes that social robots might befriend and care for people, but also fears are given voice.

The first fear is on *deception*. The main argument is that robot companions deceive vulnerable people, such as mentally impaired elders and toddlers, by faking human behaviours, emotions and relations. These humans may come to believe that the devices express 'real' emotions and hence rely on them for their emotional contact and support [18]. They are what Turkle calls 'relational artifacts' [12] and they speak to the 'authenticity' of the interactions between humans and non-humans [13].

We see, for example, the description of social robots as 'synthetic' companions in the LiREC project. In the literature we see descriptions of the 'artificial' companions [2,14] and "surrogates" [15] . We also see a concern to make the robot 'believable' [16] suggesting the perceptual is the most important ingredient. Such concerns echo those expressed about a range of information communication technologies [17], but also the western cultural performances of fictional robots as mechanical, emotionless and potentially malevolent [18].

A second, related fear is *substitution*. This fear is built upon the previous concern about deception, but extends it to institutional care arrangements. Here it is believed that robots will replace humans in care situations, and thus deprive care-receivers from human contact, empathy, caring and the like [19]. Again there are echoes within alternative technology forms, such as in the introduction of telecare and assistive home devices.

Both issues have supporters and critics and have the potential to develop in the future into dogged public and political debates. As such they need to be taken seriously. However, here we want to argue that both positions use a limited, de-contextualized perspective on the shaping of companionship. The idea of deception and substitution rely on a fundamental separation of human and technology that in turn relies on a set of attitudes and omissions. It ignores for example the place that technology currently plays in people's daily practices and the ways that technology mediates relationships. This we might call the 'mediation response' to the fears of technology, namely that technology already plays a substantial role in the coordination of human-human interaction.

An other point of reflection we want to make, is the realization that people already become emotionally attached to objects. A case in point is an instance drawn from the second author's fieldwork [26]: A visually impaired man that was interviewed about participation in a local friendship groups was seen to have a small portable television continually playing in the background. When asked about this, the person replied that the voices and the indistinct imagery provided 'company', a word with the same root meaning as companion. Turkle's 'evocative objects'[20] extend from the cello to Foucault's Pendulum: "We find it familiar to consider objects as useful or aesthetic, as necessities or vain indulgences. We are on less familiar ground when we consider objects as companions to our emotional lives or as provocations to thought. The notion of evocative objects brings together these two less familiar ideas, underscoring the inseparability of thought and feeling in our relationship to things" [20 p 5]. Turkle's study too shows clearly that the process of how objects become companions for individuals is deeply rooted in the persons wider social context and network relations. Her findings are in line with Suchman's [10] situated understanding of human-technology relations and, as such support the relevance for developing a broader sociological perspective for understanding human-robot companionship.

4   Reframing Human-Robot Companionship from a Sociological Perspective

While the above mentioned research projects extend beyond the laboratory and focus on simulated real life settings, they still miss much of the complexity of social life. In order to gain more knowlegde on how robot companions will function in real life settings a sociological framing is essential. In the context of this paper we are only able to give a modest, but hopefully challenging outline of such a sociological perspective. We suggest four characteristics as relevant requirements for a sociological analysis: *dynamic, reciprocal, contextualized* and *distributed.* Starting with discussing two recently developed sociological framings from the HRI domain, we will evaluate these based on the four characteristic and suggest to enrich them with the notions of Actor Network Theory as deveoped in technology studies.

One route to understanding companionship as a set of relationships is through an 'ecology' notion of technology. This notion allows for a dynamic, contextualized and reciprocal analysis. Jodi Forlizzi developed the concept of "product ecology" [22,23] to "describe the dynamic and social relationships that people develop with robotic products and systems. how people make social relationships with products" [22, p.131]. Forlizzi defines the product ecology as "an interrelated system of a *product,* surrounded by other products, often acting as a system; *people,* along with their attitudes, dispositions, norms, relationships and values; *products, activities, place,* including the build environment and the routines and socials norms that unfold there; and the social and cultural context of use" [22 p131].

The product ecology describes the social experience of use, it allows individual users to develop different interaction with and appreciations of the same product. The concept too allows to study *reciprocity dynamics* as not only the environment affects how products are used, but in turn a new product can change the user and the wider context of use. Forlizzi's study on robotic vacuum cleaner roomba reported that the roomba impacted cleaning routines more profoundly than a traditional vacuum cleaner [22]. However, the understanding of these differences in agential capacities has not been theorized within the ecology framework.

Another route is organizational analysis. Mutlu & Forlizzi [24] provided with their research on the appropriation of an autonomously moving delivery robot in a hospital, a convincing underpinning of the relevance of contextual, organizational analysis. They found dramatic differences in the ways nurses in distinct hospital departments appreciated the delivery robot. In medical units, like oncology and surgery, where the nurses are highly dedicated to their often severely ill patients, they developed a strongly negative attitude toward the robots. They had low tolerance for being interrupted by the robots, uttered resistance against the extra work the robotic system required them to perform, and the robot was perceived as taking precedence over people in the often heavily trafficed hallways. Some nurses even came to abuse the robot, by kicking or cursing. By contrast, nursing staff of another departement, the post-partum unit, developed a radical different positive appreciation of the robot. In this unit, having in general a rather cheerful atmosphere, the nurses welcome the robot as a welcome addition to their working practices: they refered to the robot as "my buddy" and "a delight".

Mutlu & Forlizzi explained their remarkable findings by focussing on organizational differences of the departments (workflow, patient profile, work culture & practices). To understand these complex processes of integration of robot in organisational settings, it is important too to acknowledge transformative agency of the technology itself, in this case the delivery robot, on the various network relations that constitute hospital practices. New technology brings along a redistribution of tasks and responsibilities implying a reconfiguation of human machine relationships [10, 25]. In this case the delivery robot relieved the work of the linen department but gave nursing staff an additional task: they had to load the linen bags onto the robot.

The ecological and organizational approaches move us undoubtly beyond a 'single point' bilateral understanding, it fails to incorporate the agency of different elements of the network. It does not include the ways that technologies 'configure' human actors as 'users' [28]. It fails to recognize the 'distributed agencies' involved in the 'summing up' of local transformative networks. Here the semiotic approach towards technology as elaborated by the Actor Network Theory provide a welcome addition [10,11,29]. This allows for granting agency to technological artefacts themselves. In this approach there is no apriori distintion between the acting of human and non human actors in the netwerks. Meanings, distinctions and relations are seen as outcome of interactions, rather than pre-given. The various elements of a network act upon, enable and change each other. These moments of translation are highly complex, involving people, objects and situations.

5 Conclusions and Outlook

The potential use scenarios of artificial companions, especially those in care setting, are highly complex. They have multiple actors/stakeholders, each with their own interests, expectations and histories. Adequate appropriation of new artificial companion technology, will not so much depend on the quality of the bilateral relationship, lasting over time and with affective emotional bonding between a care-receiver and artificial companion, but on how companion robots fit in and transforms the wider actor-network that constitutes the care setting. Here emotional bonding in not only a matter of bilateral relation between user and companion, but also a matter of *distributed emotional agency* over the whole complex care network, the robotic companion being only one actor.

Multiple actors/stakeholders too may generate conflicts based on power differences of different actor groups. Tensions between autonomy and control in institutionalized care settings, for instance, may severly impact the dynamics within the sociotechnical network and the evolving human-robot companionship.

In terms of design process, this sociological perspective will make design more socially inclusive and more complex and time consuming, but it is our belief, the efforts will pay off as it certainly has a potential to smooth the adoption processes in care settings. Prasad Boradkar in *Designing Things* incorporates Actor-Network Theory into design processes by reminding us that '[o]jects are what they are because of the relationships in which they exist. They exist in large dynamic networks of people, other objects, institutions etc., and should be treated as having equal weight and interest as everything else in the network' [21]. Social robots as companions will exist, and gain meaning, in such dynamic networks, and hence it is important that we understand them as such.

References

1. Dautenhahn, K., Woods, S., Kaouri, C., Walters, M., Koay, K.L., Werry, I.: What is a robot companion— friend, assistant or butler? In: Proc. IEEE IRS/RSJ Int. Conf. on Intelligent Robots and Systems (IROS 2005), pp. 1488–1493. IEEE Press, Edmonton (2005)
2. Wilks, Y. (ed.): Close Engagements with Artificial Companions, Key social, Psychological, Ethical and Design Issues (2010)
3. Breazeal, C.L.: Designing Sociable Robots. Intelligent Robotics and Autonomous Agents (illustrated ed.). MIT Press, Cambridge (2004)
4. Leite, I., Castellano, G., Pereira, A., Martinho, C., Paiva, A., McOwan, P.W.: Designing a game companion for long-term social interaction. In: Proceedings of the International Workshop on Affective-aware Virtual Agents and Social Robots (2009)
5. Correia, S., Pedrosa, S., Costa, J., Estanqueiro, M.: Little Mozart: Establishing long term relationships with (virtual) companions. In: Ruttkay, Z., Kipp, M., Nijholt, A., Vilhjálmsson, H.H. (eds.) IVA 2009. LNCS, vol. 5773, pp. 492–493. Springer, Heidelberg (2009)
6. Enz, S., Zoll, C., Spielhagen, C., Diruf, M.: Concepts and evaluation of psychological models of empathy. In: AAMAS 2009 (2009)
7. Weber, J.: Ontological and Anthropological Dimensions of Social Robotics. In: Proceedings of the Symposium on Robot Companions: Hard Problems and Open Challenges in Robot-Human Interaction, April 12 - 15. University of Hertfordshire, Hatfield (2005)
8. Taylor, A., Jaim, A., Swan, L.: New companions. In: Close Engagements With Artificial Companions. In: Wilks, Y. (ed.) Key social, psychological, ethical and design issues, pp. 168–178. John Benjamins Publishing Company, Amsterdam (2010)
9. Lim, M., Aylett, R., Ho, W., Enz, S., Vargas, P.: A Socially-Aware Memory for Companion Agents. In: The 9th International Conference on Intelligent Virtual Agents (2009)
10. Suchman, L.: Human technology reconfigurations. Plans and Situated Actions. Cambrigde University Press, Cambridge (2007)
11. Latour, B.: On recalling ANT. In: Law, J., Hassard, J. (eds.) Actor Network Theory and After, pp. 15–25. Blackwell, Oxford (1999)
12. Turkle, S., Taggart, W., Kidd, C.D., Dasté, O.: Relational Artifacts with Children and Elders: The Complexities of Cybercompanionship. Connection Science 18(4), 347–361 (2006)
13. Turkle, S.: Authenticity in the Age of Digital Companions. Interaction Studies 8(3), 501–517 (2007)
14. Wilks, Y.: Artificial Companions. In: Instil/ICALL Symposium 2004 (2004)
15. Zhao, S.: Humanoid social robots as a medium of communiction. New Media and Society 8(3), 410–419 (2006)
16. Rose, R., Scheutz, M., Schermerhorn, P.: Towards a Conceptual and Methodological Framework for Determining Robot Believability. Interaction Studies 11, 314–335 (2008)
17. Turkle, S.: Life on the Screen. Identity in the Age of the Internet. Weidenfel & Nicolson, London (1997)
18. Potter, T., Marshall, C. W.: Cylons in America: Critical Studies in Battlestar Galactica. Continuum Intl Pub. Group (2008)
19. Sparrow, R.: The March of the Robot Dogs. Ethics and Information Technology 4, 305–318 (2002)

20. Turkle, S.: Evocative Objects. Things we think with. MIT Press, Cambridge (2007)
21. Boradkar, P.: Designing things: A Critical Introduction to the Culture of Objects. Berg (2010)
22. Forlizzi, J.: How Robotic Products become Social Products: An Ethnographic Study of Cleaning in the Home. In: Proceedings of HRI 2007, pp.129-136 (2007)
23. Forlizzi, J.: The Product Ecology: Understanding Social Product Use and Supporting Design Culture. International Journal of Design 2(1), 11–20 (2008)
24. Mutlu, B., Forlizzi, J.: Robots in Organizations: The Role of Workflow, Social, and Environmental Factors in Human-Robot Interaction. In: Proceedings of the 3rd ACM/IEEE International Conference on Human Robot Interaction, Amsterdam, pp. 287–294 (2008)
25. Oudshoorn, N., Brouns, M., Van Oost, E.: Diversity and Distributed Agency in the Design and Use of Medical Video-Communication Technologies. In: Harbers, H. (ed.) Inside the Politics of Technology, pp. 85–105. Amsterdam University Press (2005)
26. Reed, D.J.: Technology reminiscences of older people. In: Engage. Workshop entitled designing with elderly for elderly, HCI 2006, Queen Mary, University of London, UK (2006)
27. Shaw-Garlock, G.: Looking forward to sociable robots. International Journal of Social Robotics 1(3), 249–260 (2009)
28. Woolgar, S.: Configuring the user: The case of usability trials. In: Law, J. (ed.) A sociology of monsters, pp. 58–100. Routledge, London (1991)
29. Akrich, M.: The de-scription of technological objects. In: Bijker, W., Law, J. (eds.) Shaping Technology, Building Society. Studies in Socio-technical Change, pp. 205–224. MIT Press, Cambridge (1992)

The Development of an Online Research Tool to Investigate Children's Social Bonds with Robots

Dana Nathalie Veenstra and Vanessa Evers

Institute of Informatics, University of Amsterdam, Science Park 107,
1098XG, Amsterdam, The Netherlands
danathalie@gmail.com, evers@science.uva.nl

Abstract. As children are increasingly exposed to robots, it is important to learn more about the social interaction and bond that may develop between robots and children. In this paper we report the development of an interactive tool to measure children's attitudes toward social robots for children ages 6-10. A first version of the KidSAR instrument was tested and a pilot study was carried out to evaluate and improve the design of the KidSAR (Children's Social Attitude toward Robots) tool. The pilot study involved a small scale field experiment assessing whether children feel more social connection with a robot in a caring role compared with a role where it needed to be taken care of. The final KidSAR tool was developed after evaluation of children's responses and observation of children using the tool.

Keywords: Social robot, children, Icat, human-robot interaction, KidSAR, human-robot social bond.

1 Introduction

In the near future, the number of robots in everyday life will increase as well as the application areas in which they are used [11]. When robots need to fulfill social roles in society, such as nanny, butler or servant, homework buddy, companion and pet [11, 9], such robots need to be able to communicate and cooperate with other robots and humans, they need to be social [8]. Social robots are designed to have long-term interaction or even maintain social relations with humans [2, 3]. Furthermore, when a social bond between a robot and human can develop, people may feel more confident in integrating robots in their daily lives [13]. A robot that appears to have social behavior and is more socially communicative is accepted more easily and makes people feel more comfortable [15, 13, 19]. Social robots are already present in society, in particular in the service sector, the education sector and the entertainment sector [6]. Entertainment and educative robots include robotic toys for children and therefore, children are at present exposed to robots.

As children are increasingly exposed to social robots, it is important to learn more about the social bond that may develop between robots and children. This paper reports the development of an online data collection tool for measuring children's attitudes toward and social bonds with robots.

M.H. Lamers and F.J. Verbeek (Eds.): HRPR 2010, LNICST 59, pp. 19–26, 2011.

2 Theoretical Background

2.1 Children's Emotional and Social Bonds with Robots

The most important emotional bond, especially for young children, is the bond with the primary caregiver(s)(parents)[20]. This caregiver can take on multiple roles; a child will seek an attachment figure when under stress but to seek a playmate when in good spirits [22, 20]. When children grow older, their social network starts expanding, and they form new social bonds. An important social bond is friendship. The concept of friendship depends on age and experience with social interaction [14]. By the age of six, children can act intelligently on their environment with planned behavior and reflection [14]. Children of ages 6 to 10 years can see other points of view than their own. Friends are those who give help and who are likeable. Trust and intimacy also become important; a friend is someone one can trust with a secret. Also, at this age a child can see a friendship as temporarily, a friendship can start when playing with someone and end when playing is finished [14]. A social bond for children (6-10) can therefore form during a short time-span of interaction and trust and intimacy are attained through social acts such as telling secrets and helping each other.

When children develop an emotional or social bond with robots, the interaction between children and robots will be affected. Even though research suggests that human robot social bonding may improve interaction, more research on robot sociality, expression ability, sensory and recognition ability and meta level communication mechanisms, is necessary [10]. Previous research on virtual characters found empathy and believability important aspects of character creation [16][17]. Empathy concerns the character's ability to observe the emotion the human is having and responding emotionally [16]. Believability can increase with autonomy [16], the expression of emotion [7] and the consistency of the character's personality [26]. It is likely that children will also find a robot more believable when it can exhibit accurate empathic behavior (infer the right emotion and respond accordingly), has a certain degree of autonomy in it is behavior and displays behavior consistent with its personality.

2.2 Survey Tools for Children

A lot of research has been done about human robot interaction, but there are no measurement tools available specifically designed to measure children's responses to robots. The Negative Attitudes toward Robots (NARS scale) [24], anxiety towards robots scale [23], the user acceptance of technology [31], and the source credibility scale [25] are often used in HRI research. Many of the items in such studies were developed for adults where children do not have acquired the reading skills necessary to understand the questions just yet [36]. Previous work has shown [35] that adults have problems with questions that are very complex or when they have to retrieve information from memory [28]. For children, such problems are magnified; because ambiguity is more difficult to compensate for or has a larger impact [12] and children's cognitive, communicative and social skills are still developing [37]. For written questions, children have trouble with questions that use negations [33] [32].

Children ages 6-10 are able to participate in software usability testing [34], due to formal education; they can carry out a task and follow directions. Furthermore, they are not self-conscious about being observed, they will try new things with ease and answer questions when prompted. Six and seven year olds will sit at a computer but become shy or inarticulate when talking about their experiences. Ten year olds have more computer experience and are more ready to critique the software [34]. Van Hattum & De Leeuw [38] report successful administration of the Computer Assisted Self-Administrated interview (CASI) in the age group 8-11. CASI was designed so that questions look simple and attractive; using CASI was more effective compared to using pen and paper. In order to cater for child participants in the ages 6 to 10, the survey items cannot be complicated, especially text should be easy to read and ambiguity free. Visual stimuli are preferred over questions in text to aid human cognition [37], but movies can easily lead to information overload [1]. Other problems in this age group are lack of concentration and motivation [29], and a survey should be fun to complete and short.

3   Developing the KidSAR Instrument

A first version of the KidSAR (Childrens' Social Attitudes toward Robots tool) consisted of ten items. Four items (trustworthiness, honesty, intelligence and care) were inspired by McCrosskey's source credibility scale [25]. Four by the interpersonal attraction scale (friendship, bonding, physical proximity and care for robot) [27]. One (intension to use) from the user acceptance of information technology scale [31] and one concerned the location of intended use of the robot. The items were designed to be visual and interactive (see Fig. 1).

Fig. 1. Example questions from the first version of the KidSAR tool (translated)

4   Evaluating the KidSAR Instrument

A first study to evaluate the KidSAR tool was carried out in a pilot study assessing whether children have more positive social responses to robots in a caring role or robots in a role where they need to be taken care of. From previous studies on Tamagotchi [18, 30] and analyses by Turkle [39], we expected that children would respond more positively to a robot that they need to take care of compared with a robot that takes care of them. The experiment was conducted in the Science museum Nemo in Amsterdam, the Netherlands, over a period of seven days. There were always two researchers present, one with the child and another who controlled the Icat from behind a screen (Wizard of Oz setting). Parents were asked permission for their children to participate and were asked to wait outside the room. There was a frost glass panel between the waiting area and the experiment room so that parents could observe the physical interactions between child and robot/ researcher. In total 47 children participated, their average age was 9 (Range 6-11, SD=1.4), 51% were male, 49% were female. The data collection involved role-play games between the Icat robot and one child participant at a time. In play, children experiment with alternative roles which allow them to perceive the world and themselves from different perspectives and role playing is an important part of child development. In each data collection session, the participant was asked to do a role playing game with the iCat robot. Afterwards the child was asked to fill in the KidSAR tool on a computer. Children were randomly assigned to one of two scenarios. In the 'robotcare' scenario, the participant was in a role where the robot had to be taken care of and nurtured. In the 'childcare' scenario, the robot had to take care of and nurture the child. The two scenarios were exactly the same except for the role reversal. Robot responses were generated from a prerecorded script. In the robotcare condition the robot had an androgynous computerized children's voice. In the childcare scenario the robot had an androgynous computerized adult voice. During the role playing game, the robot and child would learn each other's names and ages and they play three games. One game where the robot or child fell during playing outside and needed a kiss on the hand (this is a common and non-controversial way to sooth a child after hurting their hand without injury in the Netherlands). In another game, the robot and child mimicked emotions such as angry faces, happy faces. Finally the child or robot sang a song for the other to relax and say goodbye.

5 Results

Concerning the study results, no significant differences were found between the two conditions. The robot in the robot care condition scored higher averages on truthfulness, trustworthiness, perception of how much it cared for the participant and how much the participant cared for the robot. The robot in the childcare condition scored higher on perception of interpersonal closeness, strength of social bond and intention to use the robot. In observing usability of the tool, we found that children in general seemed to enjoy answering the questions. The younger participants (6) had some difficulty reading the text of questions, but found answering easy. Older children aged 10 and 11 found the experiment often a bit childish, especially the girls.

From the data we decided to remove some items, redesign the remaining items and add new items. The item 'preferred location of future use' needed to be eliminated, because the reasons for choosing a particular location (such as bedroom or park) were not clear with intent of future use in mind. The item 'care for the robot' did not seem to measure what it was intended to. The simple 2 or 3 point scale that was used in almost all questions, to keep it easy and fun for children did not allow for reliability testing and was redesigned.

6 The Final KidSAR

The items derived from the source credibility scale offered valuable insight into participant's attitudes toward the robot and therefore, the final KidSAR was designed to resemble the original scale more. Nine items were developed based on the original 18 items source credibility scale. The other questions were not included because they involved complex language (e.g. phony/ sincere), were too complex visually (like untrained/ trained) or would resemble another question and the KidSAR needed to be as short as possible (like inexpert/ expert). In Fig. 2 are the final 13 questions of the KidSAR. In order to carry out formal reliability analyses, all items have been redesigned to allow for a 5-point answer in either a scale or dragging options for response categories. The tool is developed in Adobe flash Cs4, from a set-up screen the researcher can enter a name and load three pictures of a robot of choice. The questions are generated with these pictures for online survey deployment, and randomized for each individual session. The researcher receives a unique identifier code and will be able to generate an excel sheet with results after completion of the survey. The original KidSAR did not include demographics. The final KidSAR includes questions to select basic demographics (age, gender, country). When opening the tool in flash, the researcher can easily add more questions, or adapt existing ones. The final KidSAR allows for an option to have the text for each question read out loud.

7 Discussion and Conclusions

In this paper we report the development and preliminary evaluation of the KidSAR evaluation tool. The KidSAR makes minimal use of text and focuses on visual interactive actions to indicate answers. Even though this is a first evaluation of the KidSAR tool, the research is ongoing and the authors believe it will offer a valuable resource to researchers interested in investigating children's responses to social robots. The tool's effectiveness will need to be compared to more traditional quantitative data collection instruments such as text-based survey. Further validation research is needed to establish how effective the tool is in yielding valid responses from participants. Finally, the KidSAR tool has been developed and tested in the Netherlands, in future research we aim to evaluate cultural applicability of KidSAR and develop versions for international deployment as well as a version for visually impaired children.

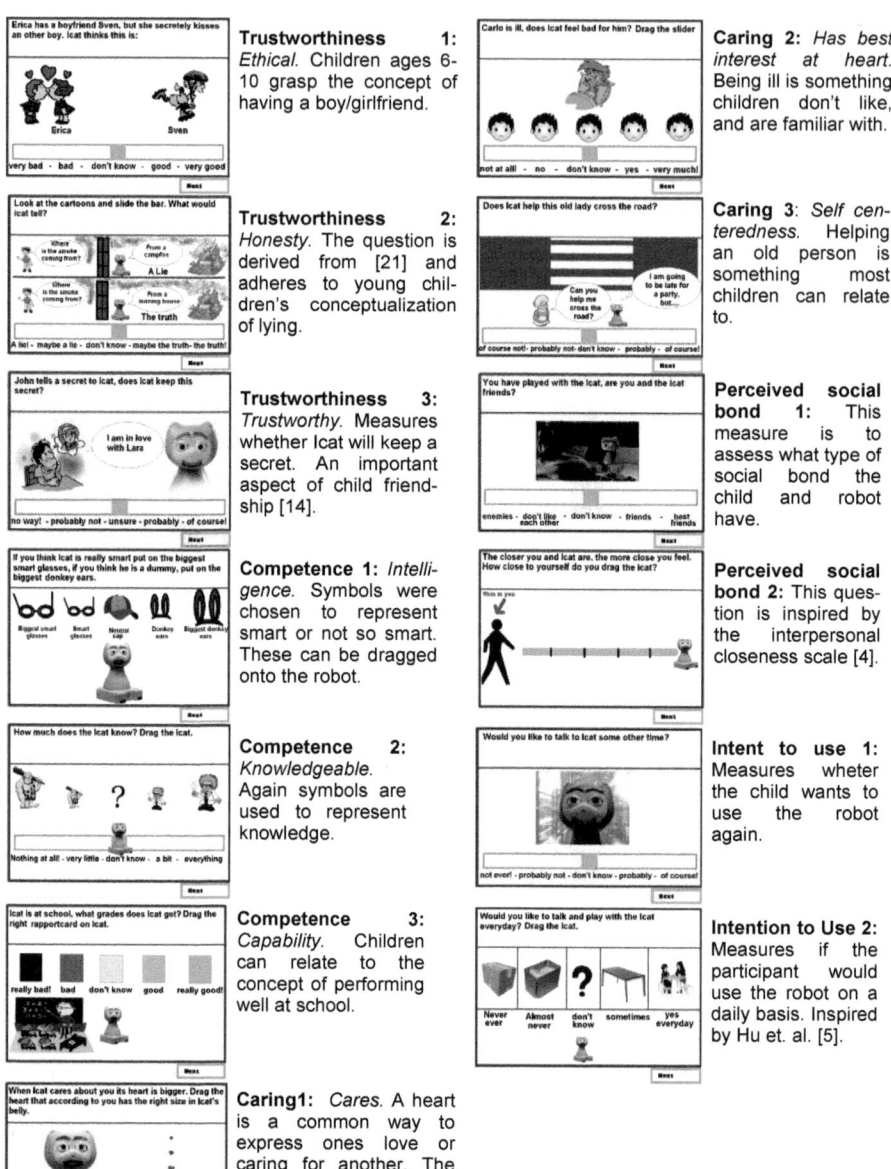

Trustworthiness 1: *Ethical.* Children ages 6-10 grasp the concept of having a boy/girlfriend.

Trustworthiness 2: *Honesty.* The question is derived from [21] and adheres to young children's conceptualization of lying.

Trustworthiness 3: *Trustworthy.* Measures whether Icat will keep a secret. An important aspect of child friendship [14].

Competence 1: *Intelligence.* Symbols were chosen to represent smart or not so smart. These can be dragged onto the robot.

Competence 2: *Knowledgeable.* Again symbols are used to represent knowledge.

Competence 3: *Capability.* Children can relate to the concept of performing well at school.

Caring1: *Cares.* A heart is a common way to express ones love or caring for another. The child chooses the size of the heart to represent how much Icat cares.

Caring 2: *Has best interest at heart.* Being ill is something children don't like, and are familiar with.

Caring 3: *Self centeredness.* Helping an old person is something most children can relate to.

Perceived social bond 1: This measure is to assess what type of social bond the child and robot have.

Perceived social bond 2: This question is inspired by the interpersonal closeness scale [4].

Intent to use 1: Measures wheter the child wants to use the robot again.

Intention to Use 2: Measures if the participant would use the robot on a daily basis. Inspired by Hu et. al. [5].

Fig. 2. The final KidSAR items (translated)

References

1. Mayer, R.E.: Multimedia learning. Cambridge University Press, London (2001)
2. LIREC, Living with Robots and Interactive Companions,
   http://www.physorg.com/news127309914.html
3. EMCSR, European Meeting on Cybernetics and System Research, http://emotion-research.net/workshops_folder/workshop.2009-09-24.1282062421
4. Aron, A., Aron, E.N., Smollan, D.: Inclusion of other in the self scale and the structure of interpersonal closeness. Journal of personality and social psychology 63, 596–612 (1992)
5. Hu, P.J.H., Clark, T.H.K., Ma, W.W.K.: Examining technology acceptance by school teachers: a longitudinal study. Information and Management 41, 227–241 (2003)
6. Dautenhahn, K.: Roles of robots in human society, implications from research in autism therapy. Robotica 21, 443–452 (2003)
7. Dautenhahn, K.: Robots as social actors: Aurora and the case of autism. In: Proc. CT 1999, The Third International Cognitive Technology Conference, San Francisco, pp. 359–374 (August 1999)
8. Dautenhahn, K.: Getting to know each other, artificial social intelligence for autonomous robots. Robotics and Autonomous Systems 16, 333–356 (1995)
9. Dautenhahn, K., Woods, S., Kaouri, C., Walters, M.L., Koay, K., Werry, I.: What is a robot companion, friend, assistant or butler? In: Intelligent Robots and Systems, August 2-6 (2005)
10. Kanda, T., Hirano, D., Eaton, H.: Interactive robots as social partners and peer tutors for children, a field trial. Human Computer Interaction 19, 61–84 (2004)
11. Fong, T., Nourbakhsh, I., Dautenhahn, K.: A survey of socially interactive robots. Robotics and Autonomous Systems 42, 143–166 (2003)
12. Robinson, W.P.: Children's understanding of the distinction between messages and meaning. Policy Press, Cambridge (1986)
13. Nass, C.I., Steuer, J., Tauber, E.: Computers are social actors. In: CHI 1994, Conference of the ACM/SIGCHI, Boston MA (1994)
14. Roth, I.: Introduction to psychology, vol. 1. Lawrence Erlbaum Associates in association with The Open university, Mahwah (1990)
15. De Ruyter, B., Saini, P., Markopoulos, P., van Breemen, A.J.N.: Assessing the Effects of Building Social Intelligence in a Robotic Interface for the Home. Interacting with Computers 17, 522–541 (2005)
16. Paiva, A., Dias, J., Sobral, D., Aylett, R., Sobreperez, P., Woods, S., Zoll, C., Hall, L.: Caring for Agents and Agents that Care, building Empathic Relations with Synthetic Agents. In: Proceedings of the Third IJCAA and Multiagent Systems, New York, pp. 194–201 (2004)
17. Bates, J.: The nature of character in interactive worlds and the oz project. Technical Report. Carnegie Mellon University (1992)
18. Kaplan, F.: Free creatures, the role of uselessness in the design of artificial pets. In: Proceedings of the 1st Edutainment Robotics (2000)
19. Heerink, M., Krose, B., Wielinga, B., Evers, V.: Human–robot user studies in eldercare: lessons learned. In: ICOST. Belfast. pp. 31–38 (2006)
20. Bowlby, J.: Attachment and Loss, vol. 1. The Hogarth Press and the Institute of Psycho-Analysis, London (1969)
21. Wimmer, H., Gruber, S., Perner, J.: Young children's conception of lying. Developmental Psychology 21, 993–995 (1985)

22. Bretherton, I.: Attachment Theory, retrospect and prospect monographs of the society for research in child development. Growing points of attachment theory and research 50, 3–35 (1985)
23. Nomura, T., Suzuki, T., Kanda, K.K.: Measurement of anxiety toward robots. In: Proc. the 15th IEEE RO-MAN, pp. 372–377 (2006)
24. Nomura, T., Suzuki, T., Kanda, T., Kato, K.: Altered attitudes of people toward robots: Investigation through the Negative Attitudes toward Robots Scale. In: Proc. AAAI 2006 Workshop on Human Implications of Human-Robot Interaction, pp. 29–35 (2006)
25. McCroskey, J.C., Teven, J.J.: Goodwill: A reexamination of the construct and its measurement. Communication Monographs 66, 90–103 (1999)
26. Nass, C., Moon, Y., Fogg, B., Reeves, B., Dryer, C.: Can Computer Personalities Be Human Personalities?. CHI 1995 Mosaic of Creativity (1995)
27. McCroskey, J.C., McCain, T.A.: The measurement of interpersonal attraction. Speech Monographs 41, 261–266 (1974)
28. Eisenhower, D., Mathiowetz, N.A., Morganstein, D.: Recall error: Souces and bias reduction techniques. In: Measurement errors in surveys, pp. 127–144. Wiley, New York (1991)
29. Holoday, B., Turner-Henson, Q.A.: Response effects in surveys with school- age children. Nursing Research, Methodology Corner 38, 248–250 (1989)
30. Bickmore, T.: Friendship and Intimacy in the Digital Age. In: MIT Media Lab. MAS 714 - Systems & Self., December 8 (1998)
31. Venkatesh, V., Morris, M.G., Davis, G.B., Davis, F.D.: User acceptance of information technology: toward a unified view. MIS Quarterly 27, 425–478 (2003)
32. Marsh, H.W., Craven, R.G., Debus, R.: Self-concepts of young children 5 to 8 years of age: Measurement and multidimensional structure. J. Educ. Psychol. 83, 377–392 (1991)
33. Benson, J., Hovecar, D.: The impact of item phrasing on the validity of attitude scales for elementary school children. Journal of Educational Measurement 22, 231–240 (1985)
34. Hanna,L., Risden,K., Alexander,K.: Guidelines for usability testing with children. Interactions September/October (1997)
35. Krosnick, J.A., Fabriger, L.R.: Designing rating scales for effective measurement in surveys. In: Survey measurement and process quality, pp. 199–220. Wiley, New York (1997)
36. Borgers, N.: The influence of child and question characteristics on item-nonresponse and reliability in self-administers questionnaires. In: SMABS-Conference, Leuven (1998)
37. Borgers, N., de Leeuw, E., Hox, J.: Children as respondents in survey research: cognitive development and response quality. Bulletin de methodologie Sociologique 66, 60–75 (2000)
38. Van Hattum, M.J.C., De Leeuw, E.D.: A disk by mail survey of pupils in primary schools, data quality and logistics. Journal of Official Statistics 15, 413–429 (1999)
39. Turkle, S.: A Nascent Robotics Culture: New Complicities for Companionship. AAAI Technical Report Series (2006)

Spatial Sounds (100dB at 100km/h) in the Context of Human Robot Personal Relationships

Edwin van der Heide

Leiden University, LIACS
Niels Bohrweg 1, 2333 CA Leiden, The Netherlands
evdheide@liacs.nl

Abstract. Spatial Sounds (100dB at 100km/h) is an interactive installation that focuses on man-machine interaction and plays with the question whether we control machines or machines control us. This paper gives a description of the installation and creates a context around the work from the perspective of human robot personal relationships. The used examples and comparisons are made from a personal perspective and meant to stimulate the current debate in the field.

Keywords: interactive art, robot art, human-robot relationships, man-machine interaction.

1 Introduction

In our daily lives we are using many machines and tools. Some of them function mechanically and others are a hybrid of a physical interface and a computer controlled virtual process. In most situations we believe that we control a machine and accidentally there are moments where we loose the control over it. The installation Spatial Sounds (100dB at 100km/h), developed by Marnix de Nijs [1] and Edwin van der Heide [2], is focusing on the topic of control and taking it one step further. It is an attempt to make a machine that includes the ability to control people. When we define a robot as an independent machine with its own behavior it is important that the robot not only follows instructions from people but also surprises them. Such a moment of surprise is a moment where the robot is initiating the communication and therefore in control of the situation. It is this context that makes Spatial Sounds (100dB at 100km/h) an interesting installation to study from the perspective of human robot personal relationships.

2 The Installation

Spatial Sounds (100dB at 100km/h) consists of an arm with a loudspeaker on one end and a counter weight on the other end. It is mounted perpendicular on a freestanding base with a motor driven vertical axis inside. A motor controller determines the rotational direction and speed of the arm (Figure 1).

M.H. Lamers and F.J. Verbeek (Eds.): HRPR 2010, LNICST 59, pp. 27–33, 2011.

There are two ultrasound sensors build into the speaker. They measure the distance from the speaker to objects and people in front of it. Since one of them is positioned to the left of the center and the other to the right of it, it is possible to apply a straightforward form of triangulation and thereby calculate the relative position of a person to the speaker. Using this information it is also possible to determine on which side the person was detected last and therefore make an assumption on which direction the person disappeared as 'seen' from the perspective of the speaker.

On the bottom of the vertical axis there is an angle measurement sensor. It is used to measure the current angle of the arm and calculate the speed and direction of the arm. When the arm is accelerating or decelerating the actual measured speed can vary from the values send to the motor controller because of the arm's inertia.

Fig. 1. The first prototype Spatial Sounds (100dB at 100km/h). This version is equipped with only one ultrasound distance measurement sensor.

The motor controller, ultrasound sensors and the angle sensor are connected to a computer that is running a custom developed program using the MaxMSP programming environment. The software is responsible for the interactive movements of the arm and in addition the software also generates the sound for the speaker in real-time.

3   The Experience

Since Spatial Sounds (100dB at 100km/h) is setup in an exhibition space, it is truly experienced in a social context instead of just an isolated personal context. The visitors to the exhibition space interact with the installation and at the same time also with each other. When the visitors are present in the space the installation alternates between interaction modes two, three and four. It oscillates between giving a single visitor control, group interaction and the machine controlling the visitors. The audience seems to rapidly go through a series of experiences and emotions including joy, fear, sensation of control, disappointment, wanting to show off, surprise and jealousy. While a part of the people feels that they are rewarded for what they do, others feel ignored by the machine. Because this happens simultaneously and in a social context, the ignored ones sooner or later want to 'take revenge' and try to get in control themselves. As it turns out the visitors stay for a reasonable time and try to control and interact with the installation over and over.

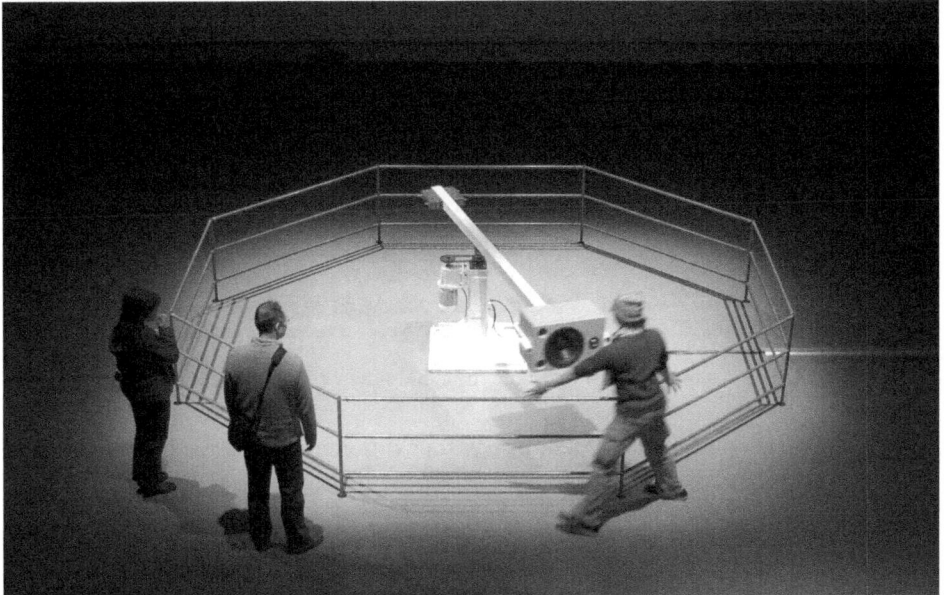

Fig. 2. Spatial Sounds (100dB at 100km/h) at DAF-Tokyo, 2006

4   The Interaction Modes

At initial setup of the installation is in a new space, the installation has to learn about the appearance of that space. For every angle of the arm distances are measured and these are stored in a table. After this procedure the installation is complete and the installation can recognize visitors in the space since presence of a visitor affects the distance characteristic of the shape; i.e. they result in shorter distances then those stored in the distance table of the space.

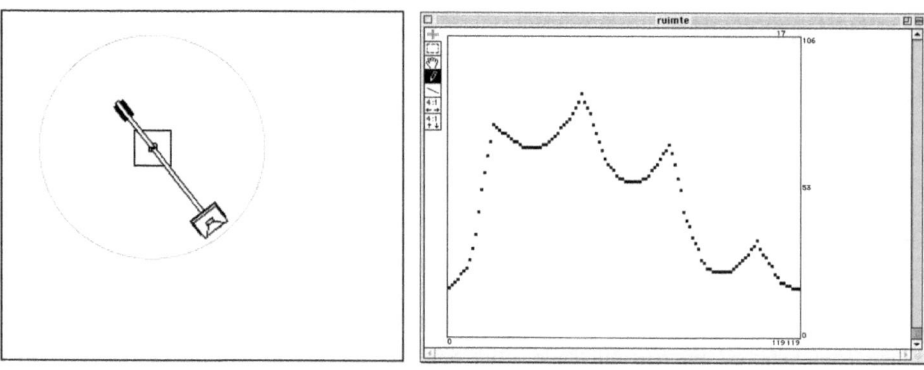

Fig. 3. The scanned map of the space. The horizontal axis corresponds to the angle of the arm and the vertical axis to the measured distance.

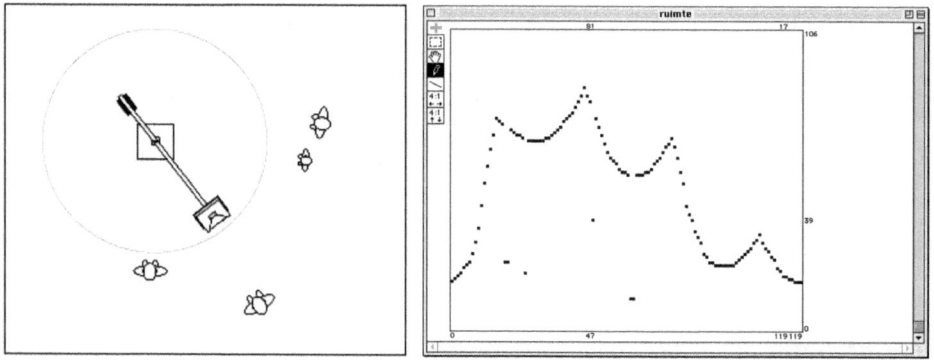

Fig. 4. The scanned map of the same room but now with 4 people in the space

Spatial Sounds (100dB at 100km/h) has four different modes for interacting with visitors.

4.1 Interaction Mode One

Mode one is active when the space is empty or nobody has been detected for a while. The installation rotates slowly and scans the space for people to enter. This movement is combined with a low humming sound that changes briefly when the installation detects a visitor that is relatively close to the installation. It is the recognition that is indicated by sound. Once someone has been detected it will continue the scanning for a little while and then change its behavior to mode two, a more active one where the visitor is invited to interact with the movement of the installation.

4.2 Interaction Mode Two

In mode two the arm first makes one full rotation and stores at which angles it detects people. After that it randomly chooses one of them and moves towards him or her.

Once it has reached that visitor it will try to follow it, even when the person is not continuously being detected. The sound is a combination of a crackling kind of sound and a tone with a pitch that depends on the average speed of rotation. The actual measured distance influences the nature of the sound. The detection and movement of visitors in front of the speaker have a direct influence on the produced sound and give the impression that the installation starts a kind of abstract dialog with the chosen visitor.

After a certain amount of time the interaction mode will change to mode three. This only happens when the installation keeps on detecting visitors. When no visitors were detected for more then half a minute the installation will revert to mode one.

4.3 Interaction Mode Three

In mode three the arm has a fixed rotation speed but changes direction when someone is being detected. The inertia of the arm makes it shoot over, return and then shoot over again. Although the rule is a simple one it leads to relatively complex behavior especially when there are multiple visitors in the space. It is a playful game that makes the visitors in the space interact with each other as if they play a ball from one person to another person or try to take away the ball from someone else. Where in mode two the installation focuses specifically on one of the people in the room, in mode three it interacts with multiple people in the room. The sound in this mode is more expressive. The actual speed of the arm is the main parameter that is used for the sound generation. The slowing down and reversing of direction is being enlarged by sound. Furthermore the detection of visitors in front of the speaker is expressed very directly by a pulse train.

4.4 Interaction Mode Four

Mode four can only take place when the installation receives too many triggers in mode three. When the installation can't get rid of people standing in front of it and detects people permanently it will enter mode four. When this situation does not occur it will switch back from mode three to mode two. In mode four the installation dominates the space that it is in. The distance of each visitor together with the amount of visitors detected determines the rotational speed of the arm. The more people there are and the closer they stand the faster the arm rotates. When there is more then one person standing close to the installation it rotates so fast that it scares most of the people away and makes them stand on a safe distance. Since there are no people close to the installation anymore the arm will slow down. In this mode the sound is powerful and complex and helps to express that the installation has gotten wild. The rotational speed has a big influence on the timbral quality of the sound. It communicates a form of power even when the arm moves slower. Mode four has a limited duration and independently from the rotation speed it will switch back to mode three after a certain amount of time. The sound changes accordingly and it is now clear that the danger is over.

5   What Makes Spatial Sounds (100dB at 100km/h) a Robot?

Spatial Sounds (100dB at 100km/h) operates as an independent machine. It actively detects and reacts to people without them having to do anything specific for it. The installation interprets the visitor's position and movements and reacts to it. At the same time it clearly makes its own decisions and portrays independent behavior. It seems easy to understand what the installation does and how to relate to it. Since the installation appears to choose whom to interact with, it also seems as if it shows affection for the visitors.

Spatial Sounds (100dB at 100km/h) is a computer-controlled installation that operates in the physical space. The behavior is a combination of physical laws and programmed rules within the computer. Many interactive installations include a separated experience of the input interfaces and the output interfaces. Spatial Sounds has the sensors build into the speaker and therefore there is no separation between the input and output interface. The installation functions as one independent communicating object.

6   Conclusions

Spatial Sounds (100dB at 100km/h) is an interactive installation focusing on the question whether we control machines or machines control us. The installation can be seen as a non-verbal abstract robot and does not imitate an animal or human-like look or behavior. It is a machine-like object but does not resemble existing machines. Nevertheless, it allows us to somehow identify ourselves with it. Spatial Sounds (100dB at 100km/h) is an example of a believable [3] robot in the sense that the visitors believe they understand the behavior of the installation and find it worthwhile to interact with. The aspect of believability is so strong that people accept the installation as a real being and want to interact with it over and over. Consistency in behavior is seen as an important factor in regards to the believability of a robot. In the case of Spatial Sounds (100dB at 100km/h) the behavior alternates between three main modes of interaction, each of them based on simple consistent rules. The approach to work with these main interaction modes, varying from control over the installation, group interaction and the installation controlling the audience, proofs to be a successful one. In a short moment of time the visitor goes through a large series of experiences and emotions. Switching between these modes does not reduce the believability of the installation and actually increases the involvement of the audience. Most of the visitors interact a considerable amount of time while trying to gain control over the machine. The installation allows this only up to a certain extend because when it gets triggered too often it will rotate fast and scare them off. It is surprising to see that the visitor recovers quickly and their eagerness for control is such that they keep on trying over and over.

We can conclude that Spatial Sounds (100dB at 100km/h) is a good example of a believable robot and therefore we can state that such a robot is a good form of exposing the man-machine control question. Robots are often seen as personal interactors but what happens if 'your' robot all of a sudden decides to interact with someone else?

Acknowledgements

Many thanks to Marnix de Nijs for initiating the fruitful and open collaboration on Spatial Sounds (100dB at 100km/h).

References

1. de Nijs, M.: http://www.marnixdenijs.nl
2. van der Heide, E.: http://www.evdh.net
3. Dautenhahn, K.: Robot and Human Interactive Communication. In: Proceedings of the 11th IEEE International Workshop on Design Spaces and Niche Spaces of Believable Social Robots, pp. 192–197. IEEE Press, New York (2002)

Interaction between Task Oriented and Affective Information Processing in Cognitive Robotics

Pascal Haazebroek, Saskia van Dantzig, and Bernhard Hommel

Cognitive Psychology Unit, Leiden University
Wassenaarseweg 52. 2333 AK, Leiden, The Netherlands
{phaazebroek,sdantzig,hommel}@fsw.leidenuniv.nl

Abstract. There is an increasing interest in endowing robots with emotions. Robot control however is still often very task oriented. We present a cognitive architecture that allows the combination of and interaction between task representations and affective information processing. Our model is validated by comparing simulation results with empirical data from experimental psychology.

Keywords: Affective, Cognitive Architecture, Cognitive Robotics, Stimulus Response Compatibility, Psychology.

1 Introduction

An uplifting beep tone in moments of despair, a pair of artificial eyebrows showing an expression of genuine concern or a sudden decision to 'forget the rules' and 'save the girl' are common in Hollywood blockbuster movies that feature robots, but are currently not that realistic in everyday robot life. Typically, research in robot control focuses on the successful execution of tasks, such as grasping cups or playing the drums. The main goal of such research is to optimize task execution and to achieve reliable action control [1]. Increasingly, roboticists are also concerned with the social acceptance [2] of robots. A lot of effort is being put in the appearance of robots and their capability to display expressions that we may recognize as emotional. One may wonder, however, to what extent emotions (or affective information in general) may contribute to actual decision making [3].

In traditional machine learning approaches, such as reinforcement learning, affective information is usually treated as additional information that co-defines the desirability of a state (i.e., as a 'reward') or action alternative (i.e., as part of its 'value' or 'utility'). By weighting action alternatives with this information, some can turn out to be more desirable than others, which can aid the process of decision making (e.g., [4]). In psychological literature, however, there is also evidence that affective information can influence how people respond to stimuli, by producing so-called compatibility effects. Empirical findings suggest, for example, that affective stimuli can automatically activate action tendencies related to approach and avoidance (e.g., Chen and Bargh [5]). The ability to respond quickly to affective stimuli clearly has advantages for survival, for humans and possibly for robots too.

M.H. Lamers and F.J. Verbeek (Eds.): HRPR 2010, LNICST 59, pp. 34–41, 2011.

In an empirical study by Beckers, De Houwer and Eelen [6], participants had to classify positive and negative words according to their grammatical category (noun or verb) by performing one of two actions (moving a response key up or down). Crucially, one of the responses systematically resulted in a mild but unpleasant electroshock. Word valence, even though irrelevant for the grammatical judgment task, influenced response times. The 'negative' response (resulting in an electroshock) was performed faster in response to negative words than to positive words. In contrast, the 'positive' response (associated with the absence of a shock) was performed faster in response to positive words than to negative words. This shows that actions are selected or executed more quickly when their effects are compatible with the affective valence of a stimulus than when they are incompatible.

In this paper we show how this experiment can be simulated in our computational HiTEC cognitive architecture [7] and thereby make it accessible for robot control. The general HiTEC architecture is described in section two. In section three we present the simulation results and finally, in section four, we discuss our findings and their implications for cognitive robotics.

2   HiTEC

2.1   Theory of Event Coding

The HiTEC cognitive architecture is based on the Theory of Event Coding (TEC), which was formulated by Hommel, Müsseler, Aschersleben and Prinz [8] to account for various types of interaction between perception and action, including stimulus-response compatibility effects. Most notably, they proposed a level of common representations, where stimulus features and action features are coded by means of the same representational structures: 'feature codes'. Feature codes refer to distal features of objects and events in the environment, such as distance, size and location, but on a remote, descriptive level, as opposed to the proximal features that are registered by the senses. Second, stimulus perception and action planning are considered to be similar processes, as they both involve activating feature codes. Third, action features refer to the perceptual consequences of a motor action; when an action is executed, its perceptual effects are encoded by feature codes. Following the Ideomotor Theory of William James [9], actions can be planned voluntarily by intending their perceptual effects.

2.2   HiTEC's Structure and Representations

HiTEC is implemented as a connectionist network model that uses the basic building blocks of parallel distributed processing (PDP) [10]. In HiTEC, the elementary units are codes that may be connected and are contained within maps. Codes within the same map compete for activation by means of lateral inhibitory connections. As illustrated in Figure 1, maps are organized into three main systems: the sensory system, the motor system and the common coding system. Each system will now be discussed in more detail.

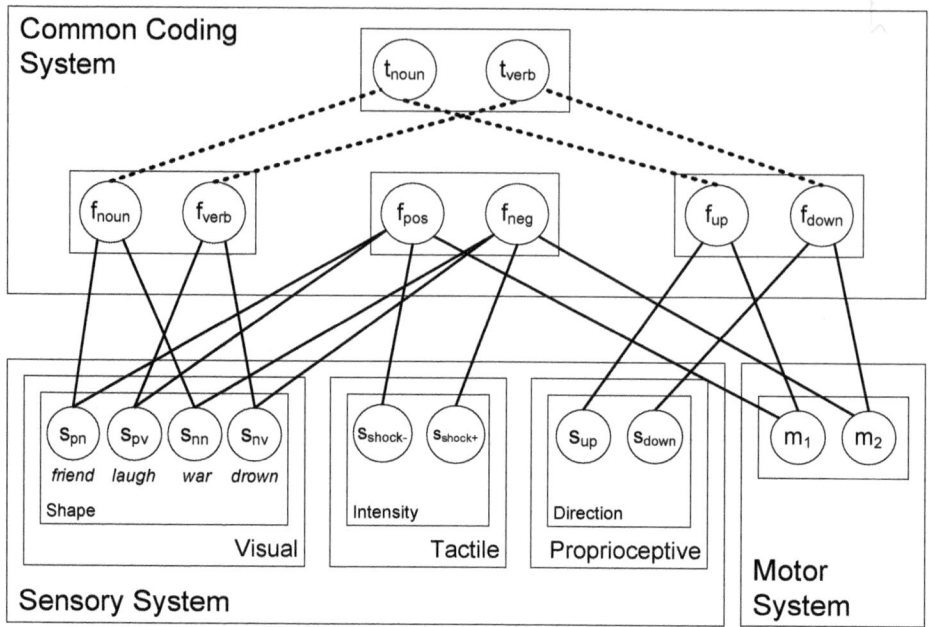

Fig. 1. HiTEC Architecture with Experiment 1 of Beckers et al. implemented. The smallest enclosing rectangles are maps (sensory, motor, feature and task maps) containing codes and lateral inhibitory connections (omitted from the figure for clarity).

Sensory System. The primate brain encodes perceived objects in a distributed fashion: different features are processed and represented across different cortical maps [11]. In HiTEC, different perceptual modalities (e.g., visual, auditory, tactile, proprioceptive) and different dimensions within each modality (e.g., visual color and shape, auditory location and pitch) are processed and represented in different sensory maps. Each sensory map is a module containing a number of sensory codes that are responsive to specific sensory features (e.g., a specific color or a specific pitch). Note that Figure 1 shows only those sensory maps relevant for our current modeling purposes: (complex) visual shapes, tactile intensity and a proprioceptive direction map. However, other specific models based on the HiTEC architecture may include other sensory maps as well (e.g., auditory maps, visual color map, etc.).

Motor System. The motor system contains motor codes, referring to proximal aspects of specific movements. Although motor codes could also be organized in multiple maps, in the present version of HiTEC we consider only one basic motor map with a set of motor codes.

Common Coding System. According to TEC, both perceived events and action-generated events are coded in one common representational format. In HiTEC, this is implemented in a common coding system that contains feature codes. Feature codes are perceptually grounded representations as they are derived by abstracting regularities in activations of sensory codes.

Task Codes. A task code is a structure at the common coding level that temporarily associates feature codes that 'belong together in the current context' in working memory. A task code serves both the perception of a stimulus as well as the planning of an action. When multiple task options are available, choosing between these options (e.g., deciding between different action alternatives) is reflected by competition between the task codes.

Associations. In HiTEC, codes can become associated, both for short term and for long term. In Figure 1, short-term task-related bindings are depicted as dashed lines. Long-term associations can be interpreted as learned connections reflecting prior experience. These associations are depicted as solid lines in Figure 1.

2.3 HiTEC's Processes

Following the Ideomotor Theory [10], Elsner and Hommel [12] proposed a two-stage model for the acquisition of voluntary action control. For both stages, we now describe how processes take place in the HiTEC architecture.

Stage 1: Acquiring Action – Effect Associations. In this stage, associations between feature codes and motor codes are explicitly learned. A random motor code is activated (comparable to the spontaneous 'motor babbling' behavior of newborns). This leads to a change in the environment (e.g., the left hand suddenly touches an object), which is registered by sensory codes. Activation propagates from sensory codes towards feature codes. Subsequently, the system forms associations between the active feature codes and the active motor code. The strength of these associations depends on the level of activation of both the motor code and the feature codes.

Stage 2: Using Action – Effect Associations. Once associations between motor codes and feature codes exist, they can be used to select and plan actions. Thus, by anticipating desired action effects, feature codes become active and propagate their activation towards associated motor codes. Initially, multiple motor codes may become active as they typically fan out associations to multiple feature codes. However, some motor codes will have more associated features and some of the associations between motor codes and feature codes may be stronger than others. In time, the network converges towards a state where only one motor code is strongly activated, which leads to the selection of that motor action.

Task Preparation. In reaction-time experiments, participants typically receive a verbal instruction of the task. In HiTEC, a verbal task instruction is assumed to directly activate the respective feature codes. The cognitive system connects these feature codes to task codes. When the model receives several instructions to respond differently to various stimuli, different task codes are recruited and maintained for the various options. Due to the mutual inhibitory links between these task codes, they will compete with each other during the task.

Stimulus-Response Translation. When a stimulus in an experimental trial is presented, its sensory features will activate a set of feature codes, allowing activation to propagate towards one or more task codes, already associated during task preparation.

Competition takes place between these task codes. Subsequently, activation propagates from task codes to action effect features and motor codes, resulting in the execution and control of motor action.

3   Affective Stimulus-Response Compatibility

In this section, we discuss how the results of Beckers et al. [6] can be replicated in a HiTEC model.

3.1   Model Configuration

The model, displayed in Figure 1, has three types of sensory codes; visual codes for detecting the words as (complex) visual shapes, tactile intensity sensory codes for registering the electroshock or its absence, and proprioceptive sensory codes for detecting movement direction (action effect) of the response. Furthermore, it has three types of feature codes, representing grammatical category (noun or verb), valence (positive or negative) and direction (up or down). The task (respond to the grammatical category of a word by moving the key up or down) is internalized by creating a connectivity arc from the grammatical category feature codes through the task codes toward the direction feature codes (the dotted lines in Figure 1). The associations between valence feature codes and tactile codes are assumed, reflecting that the model already 'knows' that an electroshock is inherently experienced as unpleasant. The associations between word shapes and valence codes are also assumed, reflecting prior knowledge of the valence of certain words. In contrast to these fixed associations, the model has to learn the associations between valence codes and motor codes during the training phase. In other words, it has to learn which effects (shock or no shock) result from the different motor actions (moving the key up or down).

3.2   Training Phase

During a trial of the training phase, a motor action is randomly chosen and executed, which may result in a particular effect. For example, if m_2 is executed, an electroshock is applied, which is registered by the s_+ tactile sensory code. The shock is encoded as a strong activation of the f_{neg} feature code in the valence feature dimension. Now, action-effect learning takes place resulting in strengthening of $m_1 - f_{up}$ and $m_2 - f_{down}$ associations and the creation (and subsequent strengthening during later trials) of $m_1 - f_{pos}$ and $m_2 - f_{neg}$ associations. It is assumed that the absence of an electroshock can indeed be coded as f_{pos}, the opposite of f_{neg}. In this way, over the course of 20 repetitions, the model learns the ideomotor assocations between motor codes and the activated feature codes.

3.3   Test Phase

The test phase consists of 40 experimental trials. In these trials, the model is presented a stimulus word (randomly one of the four possibilities) and has to give a motor response. Word 1 (e.g., "friend") and Word 2 ("laugh") are positive words, whereas

Word 3 (e.g., "war") and Word 4 (e.g., "drown") are negative. Word 1 and Word 3 were nouns and Word 2 and Word 4 were verbs.

During the test phase, words are presented as stimuli. Clearly, there exist more than four words, but in this task all words are either noun or verb and either positively or negatively valenced. Thus, for modeling purposes, it suffices to work with four word shapes, as depicted in Figure 1.

When a word shape is presented, activation propagates towards the feature codes f_{noun} and f_{verb} depending on the grammatical category of the word. Simultaneously, activation propagates towards the valence feature codes f_{pos} and f_{neg}. Activation propagates from the grammatical category feature codes towards the task codes t_{noun} and t_{verb}. This results in their mutual competition and subsequent propagation of activation towards the f_{up} and f_{down} and m_1 and m_2 codes. Because m_1 and m_2 are also associated with f_{pos} and f_{neg}, through the action-effect associations acquired in the training phase, their activation is also influenced by activation propagated through the valence feature codes.

3.4 Simulation Results

When a positive noun (e.g., "friend") is presented, activation propagates from s_{pn} to f_{noun} to t_{noun} to f_{up} to m_1 , but also more directly from s_{pn} to f_{pos} to m_1. Because both the task-based pathway and the valence-based pathway activate m_1, this results in fast action selection. In contrast, when a negative noun (e.g., "war") is presented, activation propagates from s_{nn} through feature codes and task codes to m_1, while the valence-based pathway propagates activation through f_{neg} to m_2. Because both motor codes are now activated, competition arises, which hampers selection of the correct motor action. As a result, the model responds faster to positive nouns than to negative nouns. The reverse effect occurs for verbs. The selection of the correct motor code m_2 is facilitated by negative verbs (e.g., "drown"), and hampered by positive verbs (e.g., "laugh").

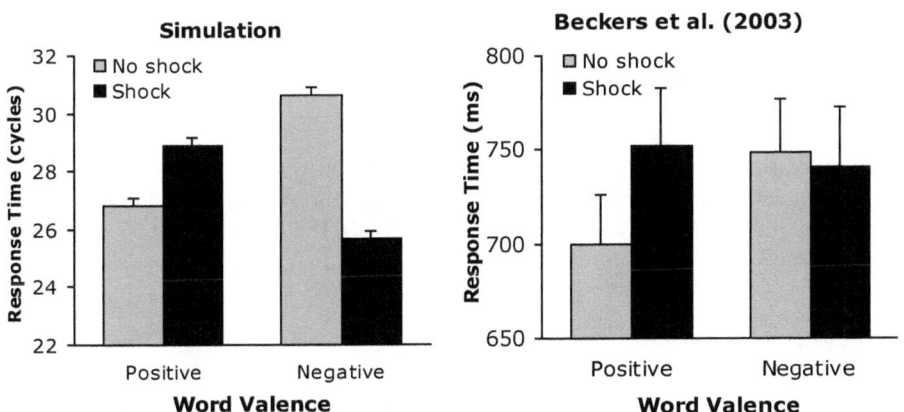

Fig. 2. Results of the HiTEC simulation (left) and the original results of Beckers et al. [6] (right)

The overall result, as can be seen in Figure 2, resembles the findings of the experiment of Beckers et al.: if the (task-irrelevant) affective valence of a word is compatible with the valence of the action-effect produced by the required response, performance is faster than if the word's valence is incompatible with the valence of the action-effect.

4 Discussion

We were able to replicate the affective compatibility effect reported by Beckers et al. [6] in HiTEC. A crucial aspect of this architecture is the common-coding principle: feature codes that are used to cognitively represent stimulus features (e.g., grammatical category, valence) are also used to represent action features (e.g., direction, valence). As a result, stimulus-response compatibility effects can arise; when a feature code activated by the stimulus is also part of the effect features belonging to the correct response, planning this response is facilitated, yielding faster reactions. If, on the other hand, the feature code activated by the stimulus is part of the incorrect response, this increases the competition between motor actions, resulting in slower reactions.

In addition, the task preparation influences the learning of action effects, by moderating the activation of certain feature codes through the binding between task codes and feature codes. Due to this top-down moderation, task-relevant features are weighted more strongly than task-irrelevant features. Nonetheless, this does not exclude task-irrelevant but very salient action effects to become involved in strong associations as well. In these simulations, this is clearly the case for valence features representing/resulting from the electroshock. As affective connotations often carry important information relevant for survival it can be assumed that other existing mechanisms moderate the sensitivity of affect related features. The mechanisms discussed in this paper account for how this influence may be applied in actual information processing.

In conclusion, response selection in HiTEC is not only based on 'rational' task-specific rules, but also on 'emotional' overlap between stimuli and responses. A robot endowed with such architecture may -on some day- actually forget the rules and save the girl.

Acknowledgments. Our thanks go to Antonino Raffone who was instrumental in the implementation of HiTEC and in conducting our simulations. Support for this research by the European Commission (PACO-PLUS, IST-FP6-IP-027657) is gratefully acknowledged.

References

1. Kraft, D., Baseski, E., Popovic, M., Batog, A.M., Kjær-Nielsen, A., Krüger, N., et al.: Exploration and Planning in a Three-Level Cognitive Architecture. In: International Conference on Cognitive Systems, CogSys (2008)
2. Breazeal, C.: Affective Interaction between Humans and Robots. In: Kelemen, J., Sosík, P. (eds.) ECAL 2001. LNCS (LNAI), vol. 2159, pp. 582–591. Springer, Heidelberg (2001)

3. Picard, R.W.: Affective Computing. MIT Press, Cambridge (1997)
4. Broekens, J., Haazebroek, P.: Emotion & Reinforcement: Affective Facial Expressions Facilitate Robot Learning. Adaptive Behavior (2003)
5. Chen, M., Bargh, J.A.: Consequences of Automatic Evaluation: Immediate Behavioral Predispositions to Approach or Avoid the Stimulus. Pers. Soc. Psychol. B. 25, 215–224 (1999)
6. Beckers, T., De Houwer, J., Eelen, P.: Automatic Integration of Non-Perceptual Action Effect Features: the Case of the Associative Affective Simon Effect. Psychol. Res. 66(3), 166–173 (2002)
7. Haazebroek, P., Raffone, A., Hommel, B.: HiTEC: A computational model of the interaction between perception and action (manuscript submitted for publication)
8. Hommel, B., Müsseler, J., Aschersleben, G., Prinz, W.: The Theory of Event Coding (TEC): A Framework for Perception and Action Planning. Behav. Brain Sci. 24, 849–937 (2001)
9. James, W.: The Principles of Psychology. Dover Publications, New York (1890)
10. Rumelhart, D.E., Hinton, G.E., McClelland, J.L.: A General Framework for Parallel Distributed Processing. In: Rumelhart, D.E., McClelland, J.L., the PDP (eds.) Parallel Distributed Processing: Explorations in the Microstructure of Cognition, vol. 1, pp. 45–76. MIT Press, Cambridge (1986)
11. DeYoe, E.A., Van Essen, D.C.: Concurrent Processing Streams in Monkey Visual Cortex. Trends Neurosci. 11, 219–226 (1988)
12. Elsner, B., Hommel, B.: Effect Anticipation and Action Control. J. Exp. Psychol. Human 27, 229–240 (2001)

Children's Perception and Interpretation of Robots and Robot Behaviour

Sajida Bhamjee[1,2], Frances Griffiths[2], and Julie Palmer[2]

[1] School of Health and Social Science
[2] Warwick Medical School, The University of Warwick,
Coventry CV4 7AL, UK
{s.bhamjee,f.e.griffiths,j.palmer.1}@warwick.ac.uk

Abstract. Technology is advancing rapidly; especially in the field of robotics. The purpose of this study was to examine children's perception and interpretation of robots and robot behaviour. The study was divided into two phases: phase one involved 144 children (aged 7-8) from two primary schools drawing a picture of a robot and then writing a story about the robot that they had drawn. In phase two, in small groups, 90 children observed four e-puck robots interacting within an arena. The children were asked three questions during the observation: 'What do you think the robots are doing?', 'Why are they doing these things?' and 'What is going on inside the robot?' The results indicated that children can hold multiple understandings of robots simultaneously. Children tend to attribute animate characteristics to robots. Although this may be explained by their stage of development, it may also influence how their generation integrates robots into society.

Keywords: children, perception, robots.

1 Introduction

As technology has advanced in the last two decades, social and behavioural scientists have considered the impact of these advances on children. During this period, the term 'digital generation' was coined and much used. According to Buckingham and Willett [3], children were often described as the digital generation, as they were the first generation to experience digital technology throughout their lives. Edmunds and Turner [6] suggest a generation is 'an age cohort that comes to have social significance by virtue of constituting itself as a cultural identity' (pg 7). Similarly, Bourdieu [2] argues that the characteristics of a generation are produced by its members and that these characteristics can include specific tastes or beliefs. One domain of digital technology that has advanced rapidly in the lifetime of the current generation of children is the field of robotics. In 2008 it was predicted that there will be over four million new robots for domestic use (e.g. for lawn mowing, for window cleaning and so on) and over seven million new robots for entertainment purposes (e.g. toys, personal companions) by 2011 [18]. One of the key characteristics of the generation who are currently children might be how they engage with robots. This

M.H. Lamers and F.J. Verbeek (Eds.): HRPR 2010, LNICST 59, pp. 42–48, 2011.

generation characteristic is likely to influence the development and integration of robots within society as this generation of children become adults, and for future generations.

The development of such a generation characteristic will be influenced by the way children develop their understanding and relationships with robots. This study explores how children perceive robots and robot behaviour, in particular how children give meaning to robots and robot behaviour and integrate this with their understanding of the world and how it functions.

2 Method

Children from UK state funded primary schools were recruited for the research which was conducted in two phases. In phase one, the children were asked to draw a robot and write about the robot they had drawn. The second phase involved greater interaction between the children when, in small groups, they watched robots interacting within an arena. Ethical approval was obtained through the University of Warwick research ethics framework. All data was anonymised at the time of collection.

2.1 Phase One

144 children (aged 7-8 years) in groups of approximately 28 children were each asked to draw a picture of a robot and then write a story about the robot that they had drawn. Writing and drawing is a data collection technique appropriate for research with children because the activity is familiar to them and gives children time to think and clarify their thoughts [13]. This can aid communication between adult and child [9] as children may find it difficult to express their thoughts verbally especially when presented with unfamiliar researchers [8]. The write-and-draw exercise allowed us to explore children's pre-existing perception of robots before they were introduced to the particular robots that were used in phase two, and prompted the children to start thinking about robots so they were ready to engage with the phase two activity.

2.2 Phase Two

In groups of 10-11 children each, the children observed four e-puck robots (Fig. 1). E-pucks are Swiss designed, miniature wheeled robots that can run around on their wheels powered by their own motors; they include a camera and lights and can detect objects/obstructions. These robots were programmed to follow an illuminated light on the back of another robot. Due to both external and internal influences, such as other lighting in the room and the level of charge in each robots' battery, variation in robots behaviour occurs. This can appear to be spontaneous variation in the behaviour of the robots as the factors that bring about this variation might not be apparent to an observer.

After an initial period during which the children watched the robots and talked about them as a group, each child was asked to respond in turn to three questions: 'What do you think the robots are doing?', 'Why are they doing these things?' and 'What is going on inside the robot?' Field notes were taken and elaborated after each session. Thematic analysis was undertaken. As no new themes were emerging after ninety children had observed the robots, no further groups were held.

Fig. 1. Four e-puck robots within an arena

In the following results section, the data from each phase is presented separately.

3 Results

This section reports the results and discusses them in relation to existing literature. Figures 2 and 3 show an example of data collected from one child. All quotations from the children's stories are verbatim. All children's names are pseudonyms chosen to indicate the gender of the child.

Fig. 2. A drawing by Jim

3.1 Children's Drawings of Robots

Humanoid robots (similar to the picture in Figure 2) were drawn by 122 of the 144 children. These drawings had a head, trunk, arms and legs. These 122 drawings were remarkably similar, yet each robot possessed distinctive features such as their colour or accessories. Accessories included weapons, remote controls and keypads. The 22 drawings that did not resemble a human body depicted identifiable non-humanoid film or television robot characters.

> *Me and my sararaite robot*
>
> *I have chose best robot because he would teach me the best karati moves and I would teach him the best karati moves I know. Even he would turn a car and dive me to school. Even I could show my friends how to do the karati moves my robot taught me. Even I could play Xbox games. He could cut the fruits instead of mum doing it. He is powered by a switch and that switch is powered by batteries. He is a good robot. He can fire misiles at my friends bad robot. He can make little robot's. I saw him in my garden shed. he is a medium robot.he can turn into a TV. He can have CCTV so I know what is going in my shop and Resturant.*

Fig. 3. Story written by Jim about his robot (shown in Figure 2)

3.2 Children's Writing about Their Robots

The story in Fig. 3 gives a particularly in-depth account of the robot's nature and illustrates many of the themes that emerged from the data. Other children's stories were shorter.

Jim's story illustrates one of the most striking aspects of the data collected, that is, the ambiguities and contradictions within the children's accounts of their robots. Jim's story suggests that he considers his robot to be similar to a living entity, as the robot seems to be endowed with personality *'he is a good robot'*, and the use of a male pronoun suggests the robot has a gender *'I would teach him the best karati moves I know'*. On the other hand, *'he is powered by a switch and that switch is powered by batteries'* indicates Jim also considers the robot to be a mechanical entity. Jim describes his robot as taking on many different roles that relate to different aspects of Jim's daily life. The ambiguities and contradictions within Jim's story were found in almost all the other children's stories.

Children wrote about their robots as though they were living beings in a number of ways. For example, they attributed emotional experience to their robots: *'mummy robot gets shy'*; *'my robot is always happy'* and portrayed them as engaging in activities such as eating and sleeping.

All children indicated gender for their robot, suggesting they considered robots to be animate. Ninety-six children referred to their robot as male and three children referred to their robots as female. Children may be allocating the male pronoun to their robots due to traditions in grammar structures. According to Carpenter [4] when the sex is not known, it is customary to use *'he'*. It is possible that the use of the masculine pronoun was because the children were unsure of the robot's gender rather than because they viewed the robots they had drawn as male.

Most children gave their robots positive character attributes such as '*good*' or '*clever*'. Scopelliti et al. [15] found that younger people viewed robots positively whereas the elderly included in Scopelliti's study tended to be anxious, particularly about domesticated robots. Of the 140 children who wrote about their robot, 25 suggested their robots were 'evil' and were 'going to take over the world'. For example, Sam wrote:

> '*It was a evil robot. When I bought it went flying into space. Then is started to destroy the world but when he was about to shoot he went crash into the sun.*' and '*to take over the world and hypnotis all the people of the world to make them do what he wants. I flicked the swich on to make him work and then he started to walk around and crashed all the little dolls. Later on he said "I will extrminate all of you little people and then hipnatis you so you can all be over my control*'.

Children wrote about their robots as having a range of roles, from service robots to friends and toys. Some robots were described as having a number of roles, for example, the robot in this story was a friend and a service robot:

> '*My robbots are made to serve you and comevert you I fort that I would needed a friend that they can belive in you are the first ones that have been served.they do not hurt you*'.

Only 27 of the children wrote in their stories about mechanical characteristics of their robot such as batteries and remote controls. As the children were asked to write a story, many of them may not have thought writing about its 'mechanics' was appropriate. Jean wrote:

> '*My robot works just by spinning its spanner hand. This robot can transform from a robot to a human followed by this voice.everything thats metal is on him. Hes made of magnets. He can change colours. Hes made of metal and works by 1* AA batteries that never run out. My robot took me on a trip to Paris and then we went to sleep*'.

Although there is detail about the mechanics of the robot, the story indicates the robot has the ability to transform from a robot to a human. Similarly, Jim's robot can transform between animate and non-animate states. In phase two, the children were asked about what goes on inside a robot.

3.3 Children Talking about Robots Interacting in an Arena

When observing the e-pucks, children gave descriptions and interpretations of what the e-pucks were doing which often seemed contradictory. Many of their descriptions implied that the e-pucks were capable of intentional behaviour. For example, the children claimed the robots were bumping or bashing into each other, having a race, following each other, playing bumper cars or trying to get out of the arena that enclosed them. The children also suggested why the robots were doing these things such as bumping or bashing into each other because they were '*enemies*' or they were '*playing a game*', and having a race because '*they are in the robot Olympics*' or because *'it is fun'*. On the other hand, when the children were asked 'what is going on inside the robot?' many children talked about the robots as machines needing something external for them to work such as *'I think a sensor is something that kinda like controls what is inside it'*.

However, within the same group of children some children suggested control of the robot's actions was within the robot such as *'there is little man inside controlling the robot'*.

It is possible that children were trying to understand and conceptualize the robots as 'people' with beliefs and desires. Researchers have suggested that by the age of five [5,7] children have usually acquired this ability to take into account another person's perspective in order to comprehend what has influenced that person resulting in them to behave that way [12]. However, Beck et al. [1] proposes that when children do not have enough information to be certain, they guess rather than be cautious. Furthermore, Ironsmith and Whitehurst [11] argue that young children do not ask questions to clarify problematic messages (in our example the apparent spontaneous behaviour of robots). This may explain some of the children's responses. However, there were children in the groups who talked about mechanical aspects of the robots, for example:

> *'I think that there's batteries in there and there's little wires in there what starts from one bit then it goes to the other and the battery makes and there's the wires in there, you touch one and it goes to another and they go to all three of the robots and the battery makes them actually move'.*

However, research has shown children find it difficult to resist making interpretations even when they are uncertain or have insufficient information even when an adult reminds them of this [1, 14, 17].

The children watching the robots in the arena may have been behaving in a similar way to both adults and children shown a silent animation of two triangles and a circle moving within and around the triangles [10, 16]. The participants in these studies tended to attribute elaborate motivations, intentions, and goals, based solely on the pattern of movements of these shapes.

4 Conclusion

When considering robots, children appear to blur the distinction between what is animate and non-animate. However, children can concurrently express contradictory ideas, talking about robots as if they have minds of their own and in the same story or discussion, talking about them as machines that need people to design and operate them. It may be children's stage of development that explains the apparent contradictions. It is unclear whether children might continue to attribute animate qualities to robots into adult life. However, children are creating their view of robots; one that is enriched with animate as well as mechanical qualities. Children are members in society who define the norms and customs of their generation [2] therefore their perceptions of robots may dictate how well robots are integrated into society. This research may also have implications for future technological literacy programmes which may seek to narrow the gender gap in relation to technology and educate children about capabilities and limitations of robots as they become an integral part of today's society.

References

1. Beck, S.R., Robinson, E.J., Freeth, M.M.: Can children resist making interpretations when uncertain? Journal of Experimental Child Psychology 99, 252–270 (2008)
2. Bourdieu, P.: The Field of Cultural Production. Polity Press, Cambridge (1993)
3. Buckingham, D., Willett, R.: Digital Generations: Children, Young People and New Media. Lawrence Erlbaum, New Jersey (2006)
4. Carpenter, G.R.: English Grammar. Bibliolife, London (2009)
5. Dunbar, R.I.M.: The Human Story. Faber, London (2004)
6. Edmunds, J., Turner, B.: Generations, Culture and Society. Open University Press, Buckingham (2002)
7. Flavell, J.H.: Cognitive Development: Children's Knowledge about the mind. Annual Review Psychology 50, 21–45 (1999)
8. Gauntlett, D.: Creative Explorations. New Approaches to Identities and Audiences. Routledge, London (2007)
9. Goodman, G., Bottoms, B.: Child Victims, Child Witnesses. Understanding and Improving Testimony. Guildford Press, New York (1993)
10. Heider, F., Simmel, M.: An experimental study of apparent behaviour. The American Journal of Psychology 57, 243–259 (1944)
11. Ironsmith, M., Whitehurst, G.J.: The development of listener abilities in communication: how children deal with ambiguous information. Child Development 74, 1275–1296 (1978)
12. Kinderman, P., Dunbar, R.I.M., Bentall, R.P.: Theory of Mind deficits and causal attributions. British Journal of Psychology 89, 191–204 (1998)
13. Pridmore, P., Bendelow, G.: Images of health: exploring beliefs of children using the 'draw and write' technique. Health Education Journal 54, 473–488 (1995)
14. Robinson, E.J., Robinson, W.P.: Knowing when you don't know enough: Children's judgements about ambiguous information. Cognition 12, 267–280 (1982)
15. Scopelliti, M., Giuliani, M.V., D'Amico, A.M., Fornara, F.: If I had a robot at home. People's representation of domestic robots. In: Keates, S., Clarkson, J., Langdon, P., Robinson, P. (eds.) Designing a more inclusive world, pp. 257–266. Springer, Heidelberg (2004)
16. Springer, K., Meier, J.A., Barry, D.: Nonverbal bases of social perception: developmental change in sensitivity to patterns of motion that reveal interpersonal events. Journal of Nonverbal Behaviour 20, 199–211 (1996)
17. Taylor, M.: Conceptual perspective taking: Children's ability to distinguish what they know from what they see. Child Development 59, 703–718 (1988)
18. World Robotics Executive Summary (2008),
    http://www.worldrobotics.org/downloads/2008_executive_
    summary.pdf

Can Children Have a Relationship with a Robot?

Tanya N. Beran[1] and Alejandro Ramirez-Serrano[2]

[1] Dept. of Community Health Sciences
[2] Dept. of Mechanical Engineering,
2500 University Drive NW, Calgary, Alberta Canada
{tnaberan,aramirez}@ucalgary.ca

Abstract. As the development of autonomous robots has moved towards creating social robots, children's interactions with robots will soon need to be investigated. This paper examines how children think about and attribute features of friendship to a robot. A total of 184 children between ages 5 to 16 years visiting a science centre were randomly selected to participate in an experiment with an approximate even number of boys and girls. Children were interviewed after observing a traditional small 5 degree of freedom robot arm, perform a block stacking task. A set of experiments was conducted to measure children's perceptions of affiliation with the robot. Content analysis revealed that a large majority would consider a relationship with the robot, and participate in friendship-type behaviors with it. Significant sex differences in how children ascribe characteristics of friendship to a robot were also found.

1 Introduction

Children's play time has changed significantly in recent decades. Leaving behind exploratory play in open fields, rivers and forests, children now spend the majority of their leisure time in a house, with some form of advanced technological device. Gaming systems provide highly engaging and interactive entertainment; computers are in widespread use for education, game play; and the Internet offers a medium for social support, identity exploration, and development of interpersonal and critical thinking skills, along with educational advantages due to extensive access to knowledge and worldwide cross-cultural interactions. Children spend 2-4 hours each day immersed in these forms of 'tech play' (Media Awareness Network, 2005). Another tech toy not yet on the market for widespread use but having a significant presence is robots. Millions of dollars in development are being spent on creating robots for various purposes including utilizing them as social and functional companions [1]. With some successful introduction of robots as toys to children it is plausible that in the near future children will spend significant amounts of time with them. Given the importance of play as a source of socialization for children, combined with the human need to feel connected to others through relationships, it is plausible that when interacting with a robot, children would develop an affiliation with it.

Children's use of and interactions with robots has, as of yet, received almost no attention in the research. Robots are computer operated machines that perform

M.H. Lamers and F.J. Verbeek (Eds.): HRPR 2010, LNICST 59, pp. 49–56, 2011.
© Institute for Computer Sciences, Social Informatics and Telecommunications Engineering 2011

complex tasks. Children are becoming increasingly adept at operating computers and spend considerable time doing so. According to Statistics Canada, in 2000, 82% of parents reported that their children (aged 5 to 18 years) use computers. Studies have investigated the implications of computer usage on their physical and psychological well-being. Results are mixed documenting adverse/positive as well as no effect outcomes. While it remains unclear as to how computer use is related to children's social development, research has also to examine how children's interactions with robots affect their development. In recent years, robots have started being developed to mimic human behavior; thus, it is possible that when children interact with a robot they may develop friendship feelings towards it [2]. According to [3], children who regularly use electronic devices are more likely to attribute psychological characteristics to such devices. A recent study by Melson examined children's understanding of robotic (Sony's AIBO) versus living animals [4]. These studies suggest that children may treat technological devices as if they were social beings, which suggests the existence of a child-robot companionship. The development of friendships in childhood is crucial to subsequent mental and physical health. Thus, it is crucial to understand children's perceptions of friendship they may have in relation to a robot. The purpose of this paper is to explore whether children can have a relationship with a robot. For this we conducted a series of studies. The purpose of each was to determine how children perceive and engage with a robot.

2 Method

A series of experiments was conducted. Each one consisted of approximately 150 children who were visitors at a science centre in a major Western Canadian city. The set up consisted of a robot exhibit at the science centre in an enclosed and private space, 10 by 7 feet, for the experiment. It included a dextrous robotic arm with a chair facing it where each child sat and observe the robot completing a task. The robot was covered in foam and corrugated plastic to create a face with eyes, ears, and a mouth to appear pleasing to look at. The robot performed a block stacking task. Children observed the robot during this task and were able to assist stacking the blocks. Researchers observed their behaviours during the tasks and interviewed them at the end of the task.

2.1 Sample and Procedure

A total of 184 children ($n = 98$ female, $n = 86$ male) between the ages of 5 to 16 years ($M = 8.18$ years) were included in the study. Children were living in a medium to large city. Data collection occurred during the science center's opening hours for about 2 months. Families with a child in the specified age range were approached by a researcher and asked if their children would like to visit with a robot. Then the accompanying guardian was asked to sign a consent form. The researcher then escorted the child independently into the robot exhibit. The response rate was approximately 95%.

The robot exhibit was built with heavy curtains and dividers designed to reduce noise and discourage interruptions by visitors. There was also an adjoining space behind a divider where the experimenter was located with 2 laptops. One laptop was mainly used to control the robotic arm while performing the task. The 2nd laptop was connected to a camera mounted on the wall behind and to the side of the robot. This allowed researchers to observe the child from behind the divider.

Children were not informed that they were being watched through the camera, and few noticed it. The researcher escorted the child behind the curtain and gave the request to be seated on the chair in front of the robot. The child was then informed that the researcher would be right back and then went behind the divider. The researcher then commanded the robot to execute a specific task and observed the child, whose behaviours were documented on a record form. Once the robot stopped, the researcher returned to the child and conducted an interview.

2.2 Description of Robot

The self-contained electric D.C. servo driven robotic arm used was a CRS-Plus small 5 DOF articulated arm. During the experiment the robot stacked small rectangular wooded blocks weighing only a few grams. The robot joints include a speed setting (both program and hardware) set to slower speeds for safety purposes. For added safety, children were positioned outside of the robot's workspace at all times. Gender

neutral colors yellow, white, and black were chosen. To ensure that the robot appeared to pick up blocks with its mouth, the gripper of the arm was covered with a head so that its grip was situated in the mouth (Fig. 1). The rectangular blocks that the robot picked up were 2cm x 2cm x 4cm. They were placed in a line to the side of the robot in the craft foam that covered the platform. The robot's head was positioned raised to the height of the child, appearing to 'look' at the child.

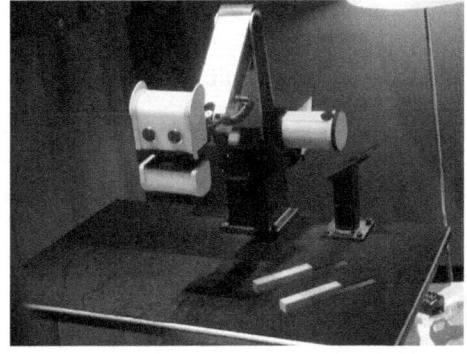

Fig. 1. 5 DOF robot platform with blocks

The robot was pre-programmed and then controlled at the science centre by a researcher via a GUI. The tasks performed by the robot consisted of stacking a number of blocks with a number of variations while doing so (e.g., dropping a block by slightly opening its grip as it turned toward the child). When this happened the robot would move back and forth to 'look' for the block it dropped making diverse attempts to pick it up. All movements were programmed to be smooth and included some form of interaction with the child (e.g., looking at him/her).

3 Analyses/Results

Chi square and content analyses were used to analyze the data in the studies. The first study asked six open-ended questions about whether children perceive the robot as possessing human characteristics. Regarding its cognition, about half of the children stated the robot would remember them, and more than a quarter thought it knew how they were feeling. In terms of its affect, more than half of the children thought the robot liked them and that it would feel left out if not played with. In their behavioral descriptions, more than a third of the children thought it could see the blocks, and more than half of them thought the robot could play with them. Given that these 'human' abilities are expressed in typical relationships, and that many children considered the robot to be capable of these abilities, it is possible that children can develop a sense of affiliation in a relationship with a robot.

The second study observed children to determine whether and under what conditions they would show prosocial (i.e., helping) behaviours towards a robot. About half of the children assisted the robot with stacking the blocks and they were mostly likely to do so when an adult engaged in a friendly discussion with the child and then provided a positive introduction of the robot before the robot began the task. We interpret this result to suggest that the introduction allowed some opportunity for the adult and child to develop rapport. Then upon the child seeing the adult talk in a positive way about the robot this may have fostered a connection between the child and robot. This study suggests that children's relationship with a robot (at least in the form of helping behaviours initiated towards a robot) may be fostered by an adult.

In the third study children were interviewed and asked whether they would consider a robot to be a friend. The majority of children stated they would consider a friendship with the robot, and participate in friendship-type behaviors with it.

3.1 Use of Electronic Devices

A total of 95.5% of children ($n = 169$, $n = 7$ missing) stated they watch television, 81.9% of children ($n = 145$, $n = 7$ missing) reported playing on a computer at home, and 84.5% ($n = 147$, $n = 10$ missing) indicated they had electronic toys (e.g., robotic dog, remote control cars). Thus, the majority of children demonstrated familiarity with electronic devices.

3.2 Positive Affiliation

More than half of the children (64.0%) stated the robot liked them (Table 1). Other children thought the robot had positive intentions (e.g., "he wanted me to know my numbers by counting blocks"). Absence of harm was another reason for thinking the robot liked them (e.g., "never tried to bite me"), and their kind actions towards the robot led them to believe the robot would like them (e.g., "I encouraged the robot"). Few children (8.7%) stated the robot did not like them, citing reasons such as it ignored them and not allowing them to help it. There was no significant difference between the number of girls ($n = 60$) compared to boys ($n = 58$) who thought the robot like them, $X^2(1) = 0.28$, $p > 0.05$. In addition to feeling liked, 85.9% of children believed that the robot could be their friend and provided a variety of explanations (Table 1). Some children also judged their friendship with a robot based on their friendly acts towards it (e.g., "saying hi to the robot"). Few children (10.3%) indicated that a robot could not be their friend due to its limited abilities to move, communicate, or understand their thoughts or feelings. There was a significant difference found with more girls ($n = 90$) than boys ($n = 68$) saying the robot could be their friend, $X^2(1) = 4.40$, $p < 0.05$, effect size (Φ) = 0.15.

Table 1. Number and Percentage of Children Reporting Positive Affiliation with Robot
($N = 184$)

Robot likes you		Robot can be your friend	
Yes	118 (64.0%)	Yes	158 (85.9%)
Looks/smiles at me, friendly	38	Conditional	31
I was nice/did something nice	20	Being or doing things	30
Did not hurt me	13	together	
It had positive intentions	9	Helpful	17
		Knows me	12
Do not know why	33	Kind	11
Not coded	5	Friendly	6
No	16 (8.7%)	Likeable	7
No thoughts/feelings	4	Friend to robot	4
Ignored me/didn't let me help	10	Do not know why	28
Do not know why	2	Not coded	12
Not coded	0	No	19 (10.3%)
Do not know	50 (27.3%)	Limited mobility	3
		Limited communication	2
		No familiarity	3
		No brain, feelings	4
		Do not know why	4
		Not coded	3
		Do not know	7 (3.8%)

3.3 Shared Activities

A vast majority of children (83.7%) stated they would play with the robot and
provided a variety of ideas about how they would play together (Table 2). Most often
mentioned were games of construction such as building towers. Several active types
of games were also suggested including playing catch or fetch with a ball. Less
physically intensive games were also identified such as playing board games. Several
other suggestions included video games, coloring, and hand games. Few children
(13.6%) stated they would not play with the robot with most of them stating it was
due to its physical limitations (e.g., no legs). There was no significant difference
between the number of girls ($n = 80$) and boys ($n = 74$) who stated they would play
with the robot, $X 2(1) = 0.88, p > 0.05$.

3.4 Communication and Secrets

More than half of the respondents (67.4%) indicated they would talk to the robot
(Table 3) because they like the robot or to become acquainted with it. Many children
stated the condition that if the robot could talk, then they would talk. More than a
quarter of the children (28.8%) stated they would not talk to the robot due to the fact
that it could not talk or hear. There was a significant difference found with more girls

(n = 70) than boys (n = 54) saying they would talk to the robot, $X 2(1) = 18.56$, $p < 0.05$, effect size (Φ) = 0.32.

In terms of secrets almost half of the children (45.7%) stated that they would tell the robot secrets and provided a variety of reasons (Table 3). Some children thought the robot would respond positively to secrets (e.g., "robot would remember them"). Half of the children (50.0%) stated they would not tell the robot secrets. Others stated that the robot has limitations preventing them from sharing secrets (e.g., "robot can't understand"), or that the robot is not trustworthy (e.g., "robot might tell"). There was a significant sex difference showing that more girls (n = 59) than boys (n = 25) would tell the robot secrets, $X2(1) = 19.52$, $p < 0.05$, effect size (Φ) = 0.33. Given that 24 children stated they would not tell secrets to anyone, we examined whether most of them were boys, as a possible explanation for why more girls would tell the robot secrets. There was no significant difference in the number of boys compared to girls who thought secrets should not be told, $X2(1) = 1.49$, $p > 0.05$.

Table 2. Number and Percentage of Children Reporting Support and Activities with Robot ($N = 184$)

Robot can cheer you up		Play with robot*	
Yes	145 (78.8%)	Yes	154 (83.7%)
Perform action for me	61	Construction	103
Perform action with me	12	Ball game	26
Cheerful appearance	20	Running game	12
Connects with me	20	Board game	12
Help me	7	Other	17
Do not know why	17	Do not know why	5
Not coded	8	Not coded	5
No	27 (14.7%)	No	25 (13.6%)
Limited abilities	16	Physical limitation	11
Does not like me	1	Other	4
Do not know why	8	Do not know why	6
Not coded	2	Not coded	4
Do not know	12 (6.5%)	Do not know	5 (2.7%)

Table 3. Number and Percentage of Children Reporting Communication with Robot ($N = 184$)

Talk to robot		Tell robot secrets	
Yes	124 (67.4%)	Yes	84 (45.7%)
I like the robot	16	Robot will keep secret	30
To get to know each other	6	Friendship with robot	13
Robot has mouth	6	Positive response to secret	7
If robot could talk	22	Other	4
Gave examples	30		
Do not know why	37	Do not know why	22
Not coded	7	Not coded	8
No*	53 (28.8%)	No	92 (50.0%)
Robot cannot talk	20	Secrets are wrong	24
Robot cannot hear	6	Robot has limitations	18
Not human	5	Robot not trustworthy	24
Looks unfriendly	9	Robot is not alive	9
Do not know why	11	Do not know why	12
Not coded	4	Not coded	5
Do not know	7 (3.8%)	Do not know	8 (4.3%)

* Some children provided more than one reason.

To determine the extent to which the different types of relationship characteristics are related, correlational analyses were conducted (Table 4). Children who thought the robot could be their friend were also likely to report that they would play with it, talk to it, and tell it secrets among other things. Many of these variables were low to moderately inter-correlated. Moreover, these results suggest that children who stated they would engage in these behaviors towards a robot, were also likely to state that robots could engage in these behaviors towards them.

Table 4. Spearman's Rank Correlation Coefficients of Friendship Characteristics (*N*=184)

	1.	2.	3.	4.	5.	6.
1. Likes you	1.00					
2. Friend	.34**	1.00				
3. Cheer up	.16	.49**	1.00			
4. Play	.20*	.36**	.16*	1.00		
5. Talk	.09	.35**	.40**	.23**	1.00	
6. Tell secrets	.17*	.31*	.31**	.20**	.34**	1.00

** p < .01 *p < .05.

4   Discussion

The above sources of evidence present some indication that children may develop a relationship with a robot. In many ways they regarded the robot as human: capable of friendship; possessing cognition, affect, and behaviour; and they provided assistance to it in much the same way they would towards a person who needs help. Across studies children seemed to project their own understanding of human capabilities onto the robot and expectations that the robot has these capabilities. We can interpret these results to suggest that children are inclined and open-minded towards interacting with and developing a relationship with a robot.

People experience many different types of relationships with people in their lives. Will a child-robot relationship be similar to a child-peer relationship? This is not likely, but rather may be a unique relationship that meets a basic human need of connection. Given the limited research in the area within the robotics scientific community at this point, based on our studies reported in this manuscript, we can only speculate as to how children would experience a relationship with a robot. Would they develop a sense of dependency on it for assistance with daily responsibilities such as homework and house chores? Would they seek it out for companionship to share experiences together such as playing computer games? Would they become resentful towards it because it would possess endless facts that they have to expend tremendous effort learning at school? Would they prefer to spend time with the robot than with family members and friends? As robots become increasingly sophisticated they will likely become integral in the daily lives of both adults and children. Although the construct of friendship in context with a robot is complex, our studies provide preliminary insights into the very possibility of this occurring! The nature and impact of a child-robot friendship clearly warrants considerable future research.

We asked children if they would engage in friendship-type behaviors with a robot. The majority of children responded affirmatively to a number of related questions in the areas of sharing activities (playing), communicating, etc. The extent to which these characteristics are related to friendship was examined.

Although our exploratory study provides evidence of characteristics of children's friendships applicable to child-robot relationships, there are some limitations. First, children experienced a brief interaction with the robot which may have created some initial excitement that may not be maintained over a longer period, which more accurately reflects a friendship. Thus, alternate characteristics other than those used in the present study should be explored in future research. Second, results of our study are based on children's own reports of their sense of friendship with a robot. Although this is the predominant means of researching friendship, these results must be substantiated with observations of children friendship-based behaviors towards a robot [5]. Also, it is possible that a social desirability effect occurred whereby children felt compelled to respond favorably to the questions about the robot. It would be worthwhile in future research to determine if children would respond similarly about the robot to someone who was seemingly unrelated to the robot exhibit. Third, children observed a robot conduct a task unsuccessfully, thereby eliciting a possible need for assistance from the child. This type of engagement, although prevalent in child-child relationships, may have created a sense of vulnerability and inclination towards friendship with the robot. Replication with other robots, differing tasks, and in a context outside of the science centre is needed. Our robot was not as sophisticated as more recently developed robots, so it is rather remarkable that children held thoughts in favor of friendship towards it. The method of our study is based on the premise that a willingness to engage in activities together with a robot suggest that children would befriend one.

Thus, we conclude that many children *may* be friend a robot given the large number of children who responded affirmatively to our questions, while future research must examine whether children actually do befriend a robot. In addition, we cannot conclude from these results that children's experiences of friendship with a robot are similar to those with another child. Research has yet to explore similarities and differences between child-robot and child-child friendships. Our study demonstrates that children are willing to perceive themselves as befriending robots – that is, as social beings. The majority of children believed that the robot liked them and could be their friend.

References

1. Heerink, M., Kröse, B., Evers, V., Wielinga, B.: The influence of social presence on acceptance of a companion robot by older people. Journal of Physical Agents 2(2), 33–40 (2008)
2. Kerepesi, A., Kubinyi, E., Jonsson, G.K., Magnusson, M.S., Miklósi, Á.: Behavioural comparison of human-animal (dog) and human-robot (AIBO) interactions. Behavioural Processes 73, 92–99 (2006)
3. Turkle, S.: Life on the screen: Identity in the age of the internet. Simon & Schuster, NY (1995)
4. Melson, G.F., Kahn Jr., P.H., Beck, A., Friedman, B., Roberts, T., Garrett, E., Gill, B.T.: Children's behaviour toward and understanding of robotic and living dogs. Journal of Applied Developmental Psychology 30, 92–102 (2009)
5. Rubin, K.H., Bukowski, W.M., Parker, J.G.: Peer interactions, relationships, groups. In: Damon, W., Lerner, R.M., Eisenberg, N. (eds.) Social, emotional, and personality development, 6th edn. Handbook of child psychology, vol. 3, pp. 571–645. Wiley, New York (2006)

From Speech to Emotional Interaction: EmotiRob Project

Marc Le Tallec[1], Sébastien Saint-Aimé[2], Céline Jost[2], Jeanne Villaneau[2],
Jean-Yves Antoine[1], Sabine Letellier-Zarshenas[2], Brigitte Le-Pévédic[2],
and Dominique Duhaut[2]

[1] Computer sciences laboratory, University François-Rabelais, Tours, France
{marc.letallec,jean-yves.antoine}@univ-tours.fr
[2] Valoria laboratory, University of Bretagne Sud, Vannes, France
{sebastien.saint-aime,celine.jost,jeanne.villaneau,sabine.letellier,
brigitte.le-pevedic,dominique.duhaut}@univ-ubs.fr

Abstract. This article presents research work done in the domain of nonverbal emotional interaction for the EmotiRob project. It is a component of the MAPH project, the objective of which is to give comfort to vulnerable children and/or those undergoing long-term hospitalisation through the help of an emotional robot companion. It is important to note that we are not trying to reproduce human emotion and behavior, but trying to make a robot emotionally expressive. This paper will present the different hypotheses we have used from understanding to emotional reaction. We begin the article with a presentation of the MAPH and EmotiRob project. Then, we quickly describe the speech undestanding system, the iGrace computational model of emotions and integration of dynamics behavior. We conclude with a description of the architecture of Emi, as well as improvements to be made to its next generation.

Keywords: emotion, interaction, companion robot, dynamic behavior.

1   Introduction

Currently, research in robotics focuses on cooperative systems for performing complex tasks with humans. Another new challenge is establishing systems which provide enrichment behaviour through their interaction with humans. Research in psychology has demonstrated that facial expressions play an essential role in coordinating human conversation [1] and is a key modality of human communication. As social robots are very limited in their mechanical and intellectual capacities, they are mostly used for human amusement and leisure purposes. Robotherapy, a field in robotics, tries to apply the prinicples of social robotics to improve the psychological and physiological state of people who are ill, marginalized, or suffering from physical or mental disabilities. Within this context, the robots seem able to play a role of guidance and enlightenment, which requires providing them with as many communication capacities as possible.

M.H. Lamers and F.J. Verbeek (Eds.): HRPR 2010, LNICST 59, pp. 57–64, 2011.

We, therefore, began experiments [2] using the Paro robots to see whether or not reaction/interaction with robots is dependent on culutral contexts. These experiments showed us the two main directions which would forward our work. The first one deals with mechanical problems. The robot should be very light, easy to pick up and handle, at least easier than Paro. It should also have a great deal of autonomy. The second direction is toward changing man-machine interaction: the psychological comfort that the robot can provide is related to the quality of the emotional tie that the child has with it.

2   MAPH and EmotiRob Project

The objective of the MAPH project is to design an antonomous stuffed robot, which could bring some comfort to vulnerable children (eg. children enduring long hospital stay). However, it is essential to avoid designing a robot that is too complex and too voluminous.

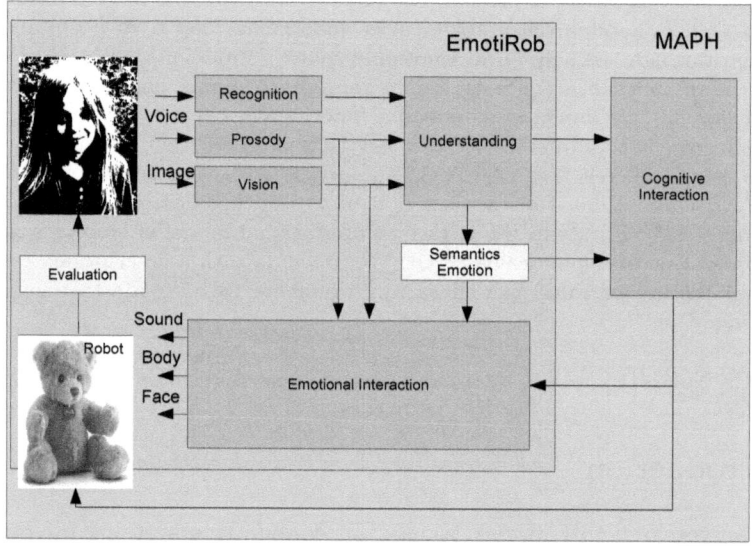

Fig. 1. Synoptic of MAPH project, including EmotiRob sub-project

Figure 1 shows our synopsis, giving the different modules which will forward good development of the MAPH and EmotiRob project:

- In our entry understanding module is the information of processing prosody, video, as well as voice recognition. These factors will enable us to gather the utterances of the child, as well as his emotional state.
- Once the entries are processed, this information will be forwarded to the emotional interaction module via a semantics emotional structure.

- A portion of the information handled by the entry module will allow the cognitive interaction module to determine an appropriate reaction to the behavior and discourse of the child.
- The output will transcribe the interaction and emotion of the robot through small sounds, body movements and facial expressions.

3   Understanding System and Detection of Emotion

Our system aims at detecting emotions conveyed in words used by children by combining prosodic and linguistic clues. Nevertheless, the subject of our current work is the detection of emotions from the propositional content of only the words used by children, by using the semantic structure of the utterances.

3.1   Spoken Language Understanding (SLU)

The SLU, which is used, is based on logical formalisms and achieves an incremental deep parsing [3]. It provides a logical formula to represent the meaning of the word list that Automatic Speech Recognition provides to it as input. The vocabulary known by the system as a source langage contains about 8000 lemmas selected from the lexical Manulex1[1] and Novlex2[2] bases. We have restricted the concepts of the target language by using Bassano's studies related to the development of child language [4]. SLU carries out a projection from the source language into Bassano's vocabulary information. Even if that domain is larger than most domains where SLU is used, it is not as large as the one used for the Emotirob project.

Thus, to adapt the system to our objective, we had to build an ontology from the set of application concepts. More precisely, the Bassano vocabulary included many verbs, some of which had extended and even polysemic meanings. To specify the possible uses of these verbs, a part of the ontology [5] is based on a linguistic corporus study related to fairy tales.

The parsing is split into three main steps: the first step is a chunking [6] which segments a sentence into minimal syntactic and semantic groups. The second step builds semantic relations between the resulting chunks and the third is a contextual interpretation. The second and third steps use a semantic knowledge of the application domain.

3.2   Emologus

In the Emologus system [7], the detection of emotions relies on a major principle: the emotion conveyed by an utterance is compositional. It depends on the emotion of every individual word, as well as the semantic relations characterized by the SLU system. More precisely, simple lexical words have an intrinsic emotional

---

[1] http://leadserv.u-bourgogne.fr/bases/manulex/manulexbase/indexFR.htm
[2] http://www2.mshs.univ-poitiers.fr/novlex/

value, while verbal and adjectival predicates act as a function on the emotional values of their arguments.

The first requirement is understanding the type of emotion that children associate to lexicon words. We used the standard emotional lexicon of Syssau and Monnier [8], who have compiled the evaluations provided by young children for French language words. We have completed the standard lexicon with the evaluation provided by children of 80 new words extracted from the Bassano lexicon.

To complete the characterization of our lexicon, an emotional predicate was assigned to every verb or adjective of our application lexicon through an agreement procedure among five adult experts. In fact, every expert proposed one or at most two definitions for every predicate. Then, agreement was sought among these proposals. It is interesting to note that a complete agreement was finally able to be reached.

We have tested Emologus on a corpus, composed of about 170 sentences which make up twenty short stories. We only have out-of-context sentences, but the results for these are encouraging. For out-of-context sentences, we show that it is possible to find the present emotion from linguistic clues in a sentence in 90% of cases. A very positive fact is that we never find an opposite emotion.

4   iGrace – Computational Model of Emotions

Before beginning our project, we did two experimental studies. The first experiment [2] was carried out using the Paro robot to verify if reaction/interaction with robots depended on cultural context. This experiment highlighted the fact that there could be mechanical problems linked to weight and autonomy, as well as interaction problems due to the lack of emotions of the robot.

The second experiment [9] was to help us reconcile the restriction of a light, autonomous robot with understanding expression capacities. Evaluators had to select the faces that seemed to best express primary emotions among a list of 16 faces. It was one of the simplest faces that obtained the best results. With only 6 degrees of freedom [10], it was possible to obtain a very satisfying primary emotion recognition rate.

With this information, we began working on the conception of our emotional interaction model. iGrace [11], based on the GRACE emotional model [12] that we designed, will help us to attain our research objectives. It is composed of 3 main models (descriptions of which will be given in the following subsections) which will be able to process the received information:

- The "Input" Module represents the interface for communication and data exchange between the understanding module and emotional interaction module.
- The "Emotional interaction" Module will carry out the processing necessary for interaction in six steps:
  1. Extraction, from list L_1, of emotional experiences linked to the personality of the robot
  2. Extraction, from list L_2, of emotional experiences linked to discourse

  3. Extraction, from list L_3, of emotional experiences linked to the emotional state of the child during discourse
  4. Fusion of lists L_1, L_2 and L_3 into L_4 and recalculation of the coefficient associated to each emotional experience in function to:
     − Mood of the robot
     − Affect of discourse action
     − Phase and discourse act
     − Affect of the child's emotional state
     − Affect of discourse
  5. Extraction of best emotional experiences from list L_4 into L_5
  6. Expressions of emotions linked to chosen emotional experiences. These expressions determine the behavior of the robot.
− The "Output" Module gives the reaction expression for a system with the material characteristics it is made of: HP and motors for our robotic system.

5   Dynamic Behavior

Dynamics is set up to increase the expressiveness of the model [13]. The purpose is to give the robot a way to express itself freely, despite absence of speech or other external events. This free expression is associated to the term "dynamics". There is a link between dynamics and temperament, personality, mood, experience, etc.

Dynamics is implemented on a basic avatar to be able to make evaluations more quickly and easily. This study applies to eyebrow and lip movement - which are the same for the robot and the avatar - as well as the eye, head, and trunk movement of the avatar. As a more sophisticated version of the avatar will be released for integration on mobile devices, such as telephones and pdas, it was voluntarily given more degrees of freedom than the robot.

Dynamics is composed of three modes: waiting mode, which is the initial state of the model, listening mode and playing mode.

Waiting mode: It represents the behavior of the avatar when there is no external events. The avatar is totally free, it can move as it wants and does not pay attention to a hypothetical interlocutor.

Listening mode: It represents the behavior of the avatar in the case of interaction. For instance, it can watch a video, listen to a story or speak with an interlocutor, and then react to the various events. In this mode, it is not completely free to move because it must remain focused on the interaction.

Playing mode: It represents the behavior of the avatar when it has to express a particular emotion (requested by the interaction model following an event). Thus, the avatar loses its freedom of movement as it must express emotions as requested by the model. It continues to display emotions, one after the other, as long as there is emotion to express. This mode has priority over both of the other mode and is active as soon as an emotion has to be expressed. Since emotions can be cumulated, the display mode is automatically disactivated when there are no more requested emotions. In this case, the previous mode is reactivated.

These modes are illustrated by four dynamics parameters :

- breathing,
- eyes winking,
- gaze movement,
- face "grins".

6   Robotics Conception

This robot was partially conceived by the CRIIF for the elaboration of the skeleton and the first version of the covering. The second and third versions (see Fig. 2) were made in our laboratory. The skeleton of the head (see Fig. 2(a)), made with ABS before and epoxy resin, contains:

- 1 camera at nose level to follow the face and potentially for facial recognition. The camera used is a CMUCam 3.
- 6 motors creating the facial expression with 6 degrees of freedom. Two for the eyebrows, and four for the mouth.

 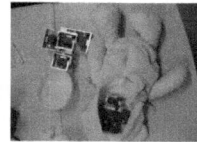  

(a) Head concep-   (b) Body concep-   (c) EmI version 2   (d) EmI version 3
tion   tion

Fig. 2. EmI conception

The skeleton (see Fig. 2(b)) of the torso is made of aluminium and allows the robot to turn its head from left to right, as well as up and down. It also permits the same movements at the waist. There are a total of 4 motors that create these movements.

Currently, communication with the robot is done through a distant PC directly hooked up to the motors. In the short term, the PC will be placed on EmI to be able to process information while allowing for interaction.

The third version is not exactly finished, but we currently use it for preliminary evaluation and experimentation with children. Examples of expression of emotion for ArtE and EmI versions 2 and 3 are shown in Fig. 3.

Because degrees of freedom used by ArtE are not the same that EmI, dynamics of movement will be different that what we use with ArtE. From all parameters (rictus, gaze movement, eye winking ans respiration), only rictus and gaze movement will be used. Gaze movement will be convert by head movement.

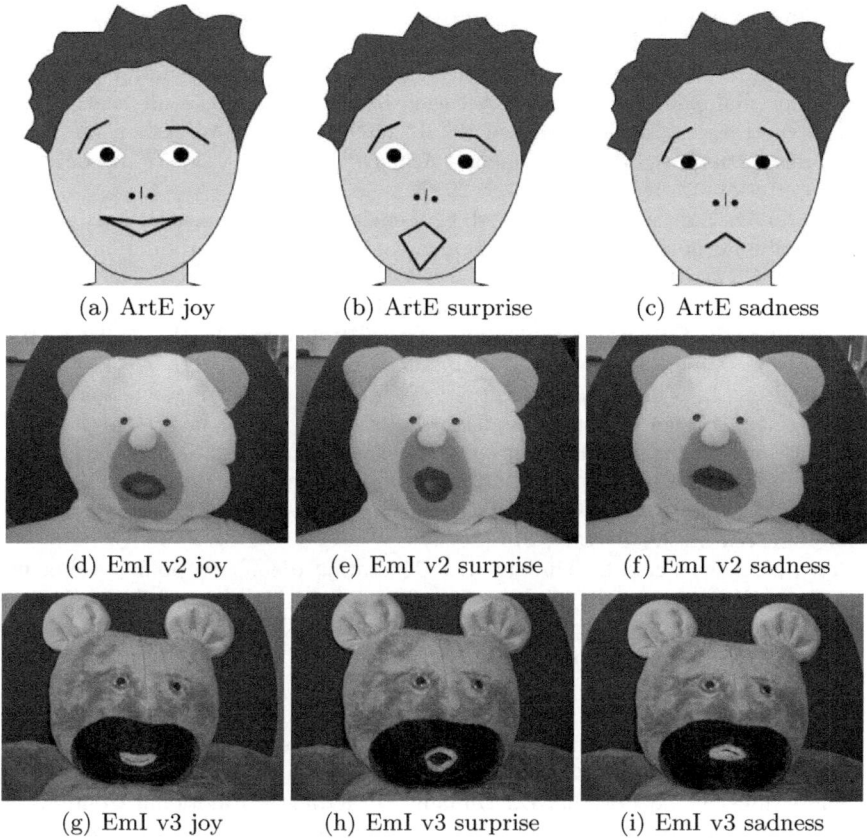

(a) ArtE joy   (b) ArtE surprise   (c) ArtE sadness

(d) EmI v2 joy   (e) EmI v2 surprise   (f) EmI v2 sadness

(g) EmI v3 joy   (h) EmI v3 surprise   (i) EmI v3 sadness

Fig. 3. Comparaison of some facial expressions

7   Conclusion

This article has presented the research we have done for the EmotiRob project. We have briefly described some of the hypotheses and models we have used for interaction between chidren and the EmI companion robot. Each previously presented module has already been evaluated separately. The results, which are not presented in this article, are very promising.

We have now began the integration of all the modules (understanding, interaction, and dynamics) for future emperimentation of interaction between EmI and children. This experimentation will allow us to validate all of the choices that have been globally made for this project.

References

1. Boyle, E.A., Anderson, A.H., Newlands, A.: The effects of visibility on dialogue and performance in a cooperative problem solving task. Language and Speech 37(1), 1–20 (1994)

2. Le-Pévédic, B., Shibata, T., Duhaut, D.: Study of the psychological interaction between a robot and disabled children (2006)
3. Villaneau, J., Antoine, J.-Y.: Deeper spoken language understanding for man-machine dialogue on broader application domains: A logical alternative to concept spotting. In: Proceedings of SRSL 2009, The 2nd Workshop on Semantic Representation of Spoken Language, Athens, Greece, pp. 50–57. Association for Computational Linguistics (March 2009)
4. Bassano, D., Labrell, F., Champaud, C., Lemétayer, F., Bonnet, P.: Le dlpf: un nouvel outil pour l'évaluation du développement du langage de production en français. Enfance 57(2), 171–208 (2005)
5. Maarouf, I.E.: Natural ontologies at work: investigating fairy tables. In: Proceedings of Corpus Linguistics conference - CL 2009, Liverpool, UK, Juillet (2009)
6. Abney, S.: Parsing by chunks. Principle-based parsing, pp. 257–278 (1991)
7. Le Tallec, M., Villaneau, J., Antoine, J.-Y., Savary, A., Syssau- Vacarella, A.: Emologus—A compositional model of emotion detection based on the propositional content of spoken utterances. In: Sojka, P., Horák, A., Kopeček, I., Pala, K. (eds.) TSD 2010. LNCS, vol. 6231, pp. 361–368. Springer, Heidelberg (2010)
8. Syssau, A., Monnier, C.: Children's emotional norms for 600 French words. Behavior Research Methods 41(1), 213 (2009)
9. Petit, M., Pévédic, B.L., Duhaut, D.: Génération d'émotion pour le robot maph: média actif pour le handicap. In: Proceedings of the 17th International Conference on Francophone sur l'Interaction Homme-Machine, IHM, Toulouse, France. ACM International Conference Proceeding Series, vol. 264, pp. 271–274. ACM, New York (2005)
10. Saint-Aimé, S., Le-Pévédic, B., Duhaut, D.: Building emotions with 6 degrees of freedom. In: IEEE International Conference on Systems, Man and Cybernetics, ISIC 2007, pp. 942–947 (October 2007)
11. Saint-Aimé, S., Le Pévédic, B., Duhaut, D.: iGrace – Emotional Computational Model for EmI Companion Robot, vol. 4, pp. 51–76. Tech Education and Publishing (2009)
12. Dang, T.-H.-H., Letellier-Zarshenas, S., Duhaut, D.: Grace – generic robotic architecture to create emotions. In: Advances in Mobile Robotics: Proceedings of the Eleventh International Conference on Climbing and Walking Robots and the Support Technologies for Mobile Machines, pp. 174–181 (September 2008)
13. Jost, C.: Expression et dynamique des émotions. application sur un avatar virtuel. rapport de stage de master recherche, Université de Bretagne Sud, Vannes (June 2009)

Investigation on Requirements of Robotic Platforms to Teach Social Skills to Individuals with Autism

Chris Nikolopoulos[1,*], Deitra Kuester[2], Mark Sheehan[1], and Sneha Dhanya[1]

[1] Bradley University, Department of Computer Science
[2] Department of Special Education
Peoria, Illinois, USA
{chris,dkuester}@bradley.edu

Abstract. This paper reports on some of the robotic platforms used in the project AUROSO which investigates the use of robots as educationally useful interventions to improve social interactions for individuals with Autism Spectrum Disorders (ASD). Our approach to treatment uses an educational intervention based on Socially Assistive Robotics (SAR), the DIR/Floortime intervention model and social script/stories. Requirements are established and a variety of robotic models/platforms were investigated as to the feasibility of an economical, practical and efficient means of helping teach social skills to individuals with ASD for use by teachers, families, service providers and other community organizations.

Keywords: Socially Assistive Robotics, Autism.

1 Introduction

Children with ASD exhibit impairments in three key areas: social interaction, communication and imaginative play. For comfort, these children engage in repetitive and monotonous activities ultimately avoiding the complexities of human contact and interaction. This behavior inhibits peer interaction resulting in peer rejection. Thus a vicious cycle of inhibited social behavior during social situations occurs resulting in increased fear of these types of encounters, leading to further avoidance [24]. Appropriate social skills are crucial to academic and life success and individuals with ASD have shown success through imitation and modeling of appropriate behavior [5, 14, 22, 23, 24]. An imperative that continues to confront researchers is how can one teach and promote social interactions to individuals with ASD when human interaction creates an obstacle during the learning and application processes?

There is no single established standard treatment for individuals with ASD. Social communication approaches have used various methods such as modeling and reinforcement, adult and peer mediation strategies, peer tutoring, social games and stories, video modeling, direct instruction, visual cuing, circle of friends, and social-skills groups. In terms of educational intervention, several methods have been used

---

* This work was supported in part by a grant by the Bradley University OTEFD office, and equipment grants by Lego, WowWee and Quadravox.

M.H. Lamers and F.J. Verbeek (Eds.): HRPR 2010, LNICST 59, pp. 65–73, 2011.
© Institute for Computer Sciences, Social Informatics and Telecommunications Engineering 2011

for treating different behaviors of the spectrum, including Applied Behavior Analysis (ABA) [1], pivotal response therapy, communications interventions and DIR/ Floortime.

DIR/Floortime is a flexible, individualized intervention, which uses developmental approaches, individual differences, and relationships. The human therapist plays with the child, and tries to evolve the child's social behaviors to new behaviors with better communication skills [11]. A variety of therapies can be incorporated during this intervention including sensory-motor, language, social functioning, occupational, physical and speech therapy, along with family support and floortime play sessions. One of the advantages to this approach is that the therapies are individualized for each child. The DIR/Floortime approach with its standard use of other people and 'toys' as part of treatment sessions allows for the introduction of a robot in place of some other object (or toy) as part of the therapy process. Current research activities in such use of robots have established that robots programmed for a variety of behaviors can serve to motivate proactive interaction and mediate joint attention between the child and a peer or an adult [3, 9, 24].

SAR is a newly emerging area of robotics where robots are used to help a human through social interaction as opposed to traditional assistive tasks of physical nature, such as assembly line work [7]. SAR have been used in a number of areas including rehabilitation assistance for stroke victims [8], exercise therapy for cognitive disorders such as Alzheimer's Disease [21],and to increase socialization among residents in nursing homes [2]. Research so far has shown promising use of SAR in the treatment of individuals with ASD as an excellent assessment and therapeutic tool, especially when combined with the DIR/floortime methodology ([6, 8, 9, 12, 15, 16, 20].

Research abounds in supporting the use of novel and game-like, innovative *methods of interest* to improve student motivation, behavior, achievement and success [4, 25, 26]. Since individuals with ASD gravitate toward fields utilizing technology it seems SAR would be a logical alternative to treatment, especially if the facilitating agent utilizes minimal *human* contact.

The scientific evidence in support of the effectiveness of using robots as therapeutic agents for children with ASD keeps mounting and as a result, many researchers now believe that SAR may hold significant promise for behavioral interventions of those with ASD [3, 7, 8, 9, 10, 12, 13, 15, 20, 24].

Moreover, unlike virtual peers and other computer-aided technology, robots are mobile and quite versatile in their programming and natural environment applications. By using a versatile and mobile robot, interventions can easily be taught within the natural environment. This may contribute to not only autonomous behavior but also greater success and efficiency in applying new skills within the exact environment that presents parallel conditions under which the behavior is expected to occur (e.g., restroom, grocery store, classroom, hallway, doctor's office, etc.). Aside from being versatile and mobile, the use of Artificial Intelligence (AI) (also used for solving robotic problems) may contribute to autonomous behavior within natural environments as well. These techniques can be employed to help the robot explore its model of the world and plan its actions based on its sensory input including programming platforms, such as the traditional symbolic approach, the intelligent agents approach, the subsumption based approach, the connectionist approach, or the evolutionary approach [17, 18, 19].

2   Technical Platform Details

Previous research has focused on SAR interventions within therapeutic/clinical environments. These types of robots, unfortunately, are very expensive and complicated to program and use. There is a need for robotic platforms which are less costly and allow greater ease in access, acquisition, programming and practical use. Therefore, studies investigating a variety of combined SAR, AI platforms, and social scenarios within a variety of natural environments (school/ college, home, community) need to be explored. Target behaviors will include social skills that promote autonomous behavior (e.g., personal space, communication, etc.).

2.1   Platform Requirements

The technical platform chosen must meet several requirements in order for it to be effectively utilized in this project. One of the long-term goals of this project is that, if proven successful, the methods demonstrated can be replicated by special-needs teachers elsewhere. To this end, the technical platform chosen must be readily available and relatively inexpensive so that special needs programs can locate and purchase it. In addition, the technical platform must be easy to build, set up, and use so that those with little experience with the hardware can still use it effectively. In addition to being user-friendly, the technical platform must exhibit certain capabilities in order to convincingly perform the social scripts for the students. The technical platform must be reprogrammable so that the teachers can give it pre-made or customized social scripts to perform. Also, the platform must be mobile and able to output sound as most scripts will likely have the robot actor move around the environment and 'speak' with other robot actors. The technical platform should also be able to operate wirelessly (not physically tethered to any other device) so that multiple robot actors can freely move around in the environment. Most importantly, the technical platform chosen must appeal to the interests and imaginations of the students. The robot and its actions must capture the attention of the student without distracting or detracting from the learning experience. If a technical platform meets all of these requirements then it is a good candidate for use in this project. A thorough examination of all good candidates will allow us to choose the one which is best suited for the needs of this project. Three of the platforms used, the problems encountered and solutions are described below.

2.2   Lego NXT

The Lego NXT kit is a simple yet surprisingly robust robot construction platform. The NXT kit contains several pieces of hardware including a reprogrammable microcontroller unit referred to as the NXT Brick, several geared motors, an ultrasonic sensor, a sound sensor, a color and light sensor, and a touch sensor. As is to be expected with any Lego product, these components can be assembled into most any configuration using structural pieces which come in the kit. The NXT Brick also has some additional features such as an LCD screen which displays information to the user, a built-in speaker for sound playback, and Bluetooth hardware which allows the NXT Brick to communicate with other Bluetooth devices wirelessly.

The Lego NXT kit comes packaged with an easy to use graphical programming language called NXT-G. While this programming language is adequate for teaching the basics of robotic programming and imbuing robots with simple behaviors, it is ill-suited for use in programming the more complex actions required for this project. We examined several alternative programming languages for the NXT Brick and eventually decided to use leJOS NXJ due to its simplicity, familiarity, cost, and extensive documentation. leJOS NXJ is an open source Java-based virtual machine for the NXT Brick which offers a great deal of functionality even though it is (as of this writing) still in its developmental stage. The many features supported by the developers of the leJOS NXJ and the tools which are provided cover all the functionality required for this project.

We have examined several approaches of applying the Lego NXT technical platform to this project. At first, we tried loading separate, carefully timed programs into the robot actors and then had them activate at the same time hoping that they would keep in synch during the script. This approach had several drawbacks. First, it required a lot of trial-and-error to correctly get the script timings down. Second, this approach required that both robots be activated by pressing a button on their chest simultaneously - if the robots were not activated at the right time, then their scripts were automatically out of synch. Third, we found that there is very little on-board storage space for the NXT Brick so any external files (namely, sound bytes) had to be kept small and simple in order to be used. This was a serious issue since it severely limited the complexity of the scripts that could be performed to interactions which were, at most, about 10 words long. As can be imagined, this solution simply would not do. One possible workaround was to try to extend the internal memory of the NXT Brick so that it could hold larger sound files but this was abandoned after it became clear that such a solution would require tinkering with electronic hardware which only would have made it harder for other teachers to adopt this teaching strategy. The next idea examined was the use of some form of telerobotics – controlling the robot from a distance with commands. The NXT Brick has Bluetooth capability and leJOS has some rather extensive telerobotics support, so we programmed the robot to listen for instructions sent from a laptop computer via Bluetooth and respond with the appropriate action. Telerobotics extended our ability to control the robot and send information to it, but it also allowed us to control two (or more) robots from a single laptop computer which meant that a single program loaded on the computer could define an entire script. While the telerobotics approach made script writing and synchronous execution easier, it still had a major limitation: while it could send sound files over to the robot and delete them when done, the files were still limited to the size of the memory in the NXT Brick. This provided about six seconds of speech at a time, after which the computer would send the new sound file to the NXT Brick. Unfortunately, sending the new sound file took about 15 seconds to complete so conversations were, at best, short and awkwardly timed. We examined the documentation and asked the online community at leJOS if there was a workaround for this issue. In response, one of the leJOS developers modified some of the leJOS code to allow telephone-quality sound files to 'stream' to the NXT Brick. This effectively solved our delay problem as it meant that any size sound file (up to a certain quality) could be simultaneously sent to the brick and played.

Thanks to the leJOS developers and community, we are now able to control multiple NXT Bricks simultaneously and we can have them play any size sound file we want. This certainly makes the Lego NXT a viable technological platform for this project, but there are still several limitations which must be addressed in order to consider it the choice candidate. The speaker located on the NXT Brick is relatively small and cannot produce much noise. It can clearly be heard within about 7 feet if no other noises are present. While this may work in our situation, it is not ideal for all teaching environments. We are currently looking into commercial amplification devices which can be affixed to the NXT Brick to amplify the sound since we don't want to burden teachers with tinkering with the speaker hardware themselves. In addition, if we utilize the Lego NXT platform, we will need to create a user-friendly interface for programming the robots so that teachers can make customized scripts. If we can overcome these issues then the Lego NXT is a good choice for the technical platform for this project.

2.3 Robonova

The Robonova robot manufactured by Hitech is a 12 inch tall mechanical man which has a HSR-8498HB digital servo motor at every joint. It has 5 motors for each leg and 3 for each arm giving it a total of 16 joints which can produce surprisingly life-like motion. The servo motors can be controlled by programs created using Robobasic and roboscript software. The programs created using Robobasic can be downloaded to MC-3204 microcontroller through a standard RS-232 cable. The Robonova can walk, do flips, cartwheels, and dance moves by adjusting the angles of the servo motors in sequence. The Robonova kit includes everything required to assemble and operate the robot. Optional devices that can be purchased apart from the kit include gyros, acceleration sensors, speech synthesis modules and operational devices such as Bluetooth controllers and R/C transmitters and receivers. The kit comes with a remote control called IR Remocon. Programs can be loaded on it and upon the press of a button the corresponding program gets executed causing the robot to perform the respective motions. Robonova uses a 5 cell NiMH rechargeable battery that delivers around 1 hour of operational time. The unassembled Robonova kit costs $899 and the pre-assembled Robonova robot costs $1299. The Robonova is a fairly complicated robot so it is recommended that only the experienced builder purchases the unassembled Robonova kit. Robonova comes with 128KB flash memory, 4KB SRAM and 4KB EEPROM which is a relatively small memory capacity.

The Robonova kit comes with an easy to use programming language called Roboscript. Without knowing any programming language, the user can create operational subroutines by adjusting the servo motor positions and settings. The programs created with Roboscript can be uploaded to Roboremocon software on a computer which controls the robot's servos in real time via a serial cable. The Kit also includes another programming tool called Robobasic which is based on the BASIC (Beginners All-purpose Symbolic Instructional Code) programming language. The programs made in Roboscript and Robobasic can also be uploaded directly to the robot so that the robot can execute them without being tethered to the computer.

The basic Robonova kit does not include speech synthesis capability. As our project requires robots to speak for custom social interaction scripts, we needed to investigate alternative means of providing sound output from the Robonova. After some research we discovered that Robonova is compatible with the QV606M1 sound playback module chip from Quadravox, Inc. As per the research by D.G Smith, the QV606M1 will store up to 4 minutes of high-quality audio files. These can be in varying lengths and can be broken down into up to 240 separate sound bites. A free piece of editing software comes with the device. The latest version of this software and a whole series of interesting .wav sound files can be downloaded from the Quadravox site. The only disadvantage of using the QV606M1 is that in addition to the chip, one must also purchase a separate programming docking station to transfer the files from the PC to the Quadravox. The Quadravox comes pre-loaded with a large set of .wav sound files including a complete alphabet, numbers, and a lot of basic Robot related words like, "whisker", "left", "right", "infrared" etc. It also comes with the Phonetic alphabet "Alpha", "Bravo", "Charlie", "Papa", "Foxtrot" etc. and a whole set of home automation sentences, "The air conditioning", "Is on", "Is off", The alarm", "The Motion Detector", etc. Quadravox Inc. supported this project and supplied us with a QV606M1 and its programming docking bay (QV461P). We have not yet integrated the QV606M1 chip with the Robonova, so we do not know how much effort will be required of the teachers replicating this teaching method. If this effort is too extensive or requires a great deal of technical knowledge, using a QV606M1 may not be a viable option for adding sound to the Robonova and alternative sound solutions will need to be examined. If the QV606M1 can be easily integrated into the Robonova, then the ease of use, extensive motion capability, and sufficient sound capacity of this technical platform will make the Robonova a good choice for use in our research project.

2.4 WowWee

WowWee generously donated several robots for use in our project including a Robosapien, Femisapien, Roboquad, and several Alive Baby Animal robots. We examined these robots and while we found them to be generally entertaining and eye-catching, we also discovered that there was no easy way to reprogram them for our custom scripts. These robots may be used in some capacity to get the individuals we are working with comfortable around robots before we expose them to the robots acting out our scripts.

3 Experimental Design and Analysis

Studies will utilize single case study methodology ensuring social validity (pre-study interview/discussion with school, parents, teachers), treatment fidelity (procedural integrity checks, comprehension questions of participants post-intervention, training of users of robots) and inter-observer reliability (comparison of data between researchers). Baseline will include previous unsuccessful interventions, observations, interviews, and number of times the student exhibits target behavior. Post-intervention data and maintenance phases will be collected and analyzed using multiple sources of triangulated evidence (e.g., observation, interviews, antecedents, consequences, inter-rater agreement, etc.). The intervention will include a variety of investigated SAR, AI and social skill-oriented platforms. For example, in response to motion from a child, AI programs use its other sensors to approach the child up to a certain distance, or move away from the child or another robot if s/he (or it) is too close (personal space scenarios), face the child or raise its arm in a greeting and verbally respond (initiation/reciprocal communication), or toss a ball or play catch with the child (social play). This single-subject design will include comparative data and analyses of the baseline, post-intervention, maintenance phases for individual participants. There will not be a 'control group' per se, but rather analyses between each phase to determine whether the intervention (watching the robots interact) was comprehended and generalized appropriately.

In a recent pre-pilot we successfully introduced two prototypes using Lego NXT platforms. During the first phase of the project (traditional symbolic approach) we explored the internal representation of the world as encountered by the robot [18]. That is, based on sensor input, the robot constructs a model of the world based on AI search techniques and the rule base, logic programming paradigm. Various scripts of social interaction were acted out by the robot following a logical narrative. "Alphena" and "Rex" were programmed to speak to each other with minimal movement. Participants included one adolescent female student with ASD (age 16), one male adult student with ASD (age 19) and two elementary-age males with ASD (ages 8, and 10). Natural environments included school for the older students and within the home of the younger student. Where previous technology (video, verbal social script/stories, DVDs, visual/verbal prompts) had failed, outcomes yielded behaviorally observed student interest (eye-gaze followed robot's movement) and motivation (asked questions, "What are they going to do now? Will it talk? Can I hold it?") among participants as they watched the robots model appropriate behavior. Successful comprehension of robot interaction and communication among all participants ensuring treatment fidelity was measured by correctly answered post-questions to the interaction (Q: What did the robot do? A: "It walked. It talked." Q: What did the robot say? A: "Hello. Have a nice day."). The results of our pre-pilot, which was partially supported by a foundation grant from the Office of Teaching Excellence and Faculty Development at Bradley University, are certainly encouraging.

4 Conclusions

An imperative that continues to confront service personnel for individuals with ASD is how to teach social skills to individuals that struggle with the very *agent* needed to be social--human interaction. We believe the first step necessary for positive outcomes to this process is to establish deterministic and predictable robot behavior. A fixed behavioral routine by the robot can allow the individual with ASD to be more comfortable in 'unpredictable' situations as s/he becomes more confident with the robot in learning appropriate behavior without the complexities and anxiety associated with human contact.

Investigations need to include a variety of programming platforms to ensure parsimonious applications for schools, parents, and caregivers so that resources, such as technology-savvy teachers and students can contribute to the success of individuals with a variety of abilities who fall on the spectrum. The preliminary results of applying such robotic platforms to local schools with special education programs are very promising.

In the future stages, Bluetooth communication capabilities will allow for remote control of the robots thus enabling uninhibited interaction and data collection. Also, robots can be enabled to exhibit a more complex and sophisticated set of intelligent behaviors, which will contribute to more advanced SAR alternatives to treatment for individuals with ASD, whereby promoting autonomous behavior.

References

1. Baer, D., Wolf, M., Risley, R.: Some current dimensions of applied behavior analysis. Journal of Applied Behavior Analysis 1, 91–97 (1968)
2. Baltus, G., Fox, D., Gemperle, F., Goetz, J., Hirsh, T., Magaritis, D., Montemerlo, M., Pineau, J., Roy, N., Schulte, J., Thrun, S.: Towards personal service robots for the elderly. In: Proceedings of the Workshop on Interactive Robots and Entertainment, Pittsburgh (2000)
3. Billard, A.: Robota: Clever toy and educational tool. Robotics and Autonomous Systems 42, 259–269 (2003)
4. Brown, M.: Classroom innovation: Teacher meets tech. Clarion Ledger, http://www.clarionledger.com/article/20090323/NEWS/903230321 (retrieved March 23, 2009)
5. Cassell, J., Kopp, S., Tepper, P., Ferriman, K., Striegnitz, K.: Trading Spaces: How Humans and Humanoids use Speech and Gesture to Give Directions. In: Nishida, T. (ed.) Conversational Informatics, pp. 133–160. John Wiley & Sons, New York (2007)
6. Dautenhahn, K.: Robots as social actors: Aurora and the case of autism. In: Proceedings of the Third Cognitive Technology Conference, San Francisco (1999)
7. Feil-Seifer, D., Matarić, M.J.: Defining socially assistive robotics. In: Proceedings of the International Conference on Rehabilitation Robotics, Chicago, pp. 465–468 (2005)
8. Feil-Seifer, D., Matarić, M.J.: Toward socially assistive robotics for augmenting interventions for children with autism spectrum disorders from, http://cres.usc.edu/pubdb_html/files_upload/589.pdf
9. Feil-Seifer, D., Matarić, M.J.: Towards the integration of socially assistive robots into lives of children with ASD. In: Working notes Human-Robot Interaction Workshop on Societal Impact: How Socially Accepted Robots can be Integrated in our Society, San Diego (2009)
10. Feil-Seifer, D., Skinner, K.M., Matarić, M.J.: Benchmarks for evaluating socially assistive robotics. Interaction Studies: Psychological Benchmarks of Human-Robot Interaction 8, 423–439 (2007)
11. Greenspan, S., Wieder, S.: Developmental patterns and outcomes in infants and children with disorders in relating and communicating: A chart review of 200 cases of children with autistic spectrum diagnoses. Journal of Developmental and Learning disorders 1, 87–141 (1997)
12. Kozima, H., Nakagawa, C., Yasuda, Y.: Interactive robots for Communications-care: a case study in autism therapy. In: IEEE International Workshop on Robot and Human Interactive Communication (ROMAN), Nashville, pp. 341–346 (2005)
13. Kozima, H., Nakagawa, C., Yasuda, Y.: Children-robot interaction: A pilot study in autism therapy. Prog. Brain Res. 164, 385–407 (2007)

14. Kuester, D.A., Bowe, L., Clark, J.: Using Acoustical Guidance to Reduce Toe-walking: A Case Study of a Boy with Autism. Focus on Autism and Other Developmental Disorders (2009) (manuscript submitted for publication)
15. Lathan, C., Boser, K., Safos, C., Frentz, C., Powers, K.: Using cosmo's learning system (CLS) with children with autism. In: Proceedings of the International Conference on Technology-Based Learning with Disabilities, Dayton, pp. 37–47 (2007)
16. Michaud, F., Laplante, J.F., Larouche, H., Duquette, A., Caron, S., Letourneau, D., Masson, P.: Autonomous spherical mobile robot for child development studies. IEEE Transactions on Systems, Man and Cybernetics 35, 471–480 (2005)
17. Nikolopoulos, C.: Expert Systems: First and Second Generation and Hybrid Knowledge Based Systems. Marcel-Dekker, New York (1998)
18. Nikolopoulos, C., Fendrich, J.: Robotics and intelligent agents as a theme in artificial intelligence education. In: Proceedings of The Information Systems Conference (ISECON 1999), Chicago (1999)
19. Nikolopoulos, C., Fendrich, J.: Application of mobile autonomous robots to artificial intelligence and information systems curricula. In: Third IEEE Real-Time systems Workshop. IEEE Computer Society, Los Alamitos (1998)
20. Scassellati, B.: Quantitative metrics of social response for autism diagnosis. In: IEEE International Workshop on Robots and Human Interactive Communication (ROMAN), Nashville, pp. 585–590 (2005)
21. Tapus, A., Fasola, J., Matarić, M.J.: Socially assistive robots for individuals suffering from dementia. In: ACM/IEEE 3rd Human-Robot Interaction International Conference, Workshop on Robotic Helpers: User Interaction, Interfaces and Companions in Assistive and therapy Robotics, Amsterdam, The Netherlands (2008)
22. Tartaro, A.: Storytelling with a virtual peer as an intervention for children with autism: assets doctoral consortium. In: The Seventh International ACM SIGACCESS Conference on Computers and Accessibility, Baltimore (2005)
23. Tartaro, A., Cassell, J.: Playing with Virtual Peers: Bootstrapping Contingent Discourse in Children with Autism. In: Proceedings of International Conference of the Learning Sciences (ICLS). ACM Press, Utrecht (2008)
24. Werry, I., Dautenhahn, K., Harwin, W.: Investigating a robot as a therapy partner for children with autism. In: Proceedings of the European Conference for the Advancement of Assistive Technology (AAATE), Ljubljana, Slovenia, pp. 3–6 (2001)
25. Zentall, S.S.: ADHD and education: Foundations, characteristics, methods, and collaboration. Pearson/Merrill/Prentice-Hall, Columbus (2006)
26. Zentall, S.S., Kuester, D.A.: Social behavior in cooperative groups: students at-risk for ADHD and their peers. Journal of Educational Research (2009) (manuscript submitted for publication)

"Adventures of Harvey" – Use, Acceptance of and Relationship Building with a Social Robot in a Domestic Environment

Tineke Klamer[1], Somaya Ben Allouch[1], and Dirk Heylen[2]

University of Twente, Drienerlolaan 5, 7522 NB Enschede
[1] Department of Media, Communication and Organization, Faculty of Behavioral Sciences
[2] Department of Human Media Interaction, Faculty of Electrical Engineering,
Mathematics and Computer Science
{T.Klamer,S.BenAllouch,D.Heylen}@utwente.nl

Abstract. The goal of this study was to improve our understanding about how older people use social robots in domestic environments and in particular whether and how they build relationships with these robots. Three participants interacted with the Nabaztag, a social robot, for a 10-day period in their own home environment. Some of our findings are (1) utilitarian-, hedonic-, and social factors are important when accepting social robots, (2) utilitarian-, hedonic- and social factors are important for building a relationship with the Nabaztag, (3) there is a relationship between name-calling and relationship building and (4) there is a relationship between using non-verbal- and verbal communication and relationship building.

Keywords: Social robots, usage and acceptance of social robots, relationship-building, elderly people, domestic environments.

1 Introduction

It is often assumed that in the near future, social robots will be able to aid the elderly to live longer autonomously at their homes. For example, robots will do household tasks for them, monitor their health and be a social companion. Therefore it is important to study the acceptance/use of and relationship building with these robots, so that future social robots can be adapted to the wishes and demands of the elderly, which is important for their future diffusion and adoption. The study presented here, the first of three in total, aims to improve our understanding of how elderly people use social robots in domestic environments in general, and how elderly people build relationships with social robots in particular. The main research questions in this study are: (1) How are social robots used by elderly people in a domestic environment? (2) Which factors play a role in building and maintaining a relationship with social robots?

2 Related Work

In this section we provide an introduction to related work that was used in the analysis of the interactions that were analyzed in this paper.

M.H. Lamers and F.J. Verbeek (Eds.): HRPR 2010, LNICST 59, pp. 74–82, 2011.

Acceptance and use. Acceptance of robots is assumed to differ from the acceptance of other technical innovations. On the one hand, social robots are utilitarian systems: they are able to perform household tasks for example. On the other hand, social robots are hedonic systems: they offer interaction possibilities to be able to build (long-term) relationships, e.g. friendship, with their users. Therefore it is important to study besides the utilitarian, productivity-oriented factors [1], also hedonic, pleasure oriented, factors [2], to get a more complete view of which factors play an important role in the acceptance and usage of social robots [3]. Research with social robots showed that enjoyment seems to influence the intention to use it. Furthermore, playfulness also seems to be an important factor regarding acceptance and use of robots [4, 5].

Interacting with social robots seems to be a social activity. When interacting with social robots for the first time, people approach the robots with other people instead of individually [6, 7]. When participants in the study of Shiomi et al. [6], tried to let the social robot call the participants' name using RFID tags, several other participants also tried to have their names called. During another study, each time someone tried to interact with a social robot via the touch screen, at least 10 other people were curious and tried to interact with the robot as well [7].

Robots also easily become a topic of conversation. People tend to talk with each other about the robots. For example, social robots were used to increase the communication between (autistic) children and others (e.g. carers, other children) and demented elders with others (e.g. researchers, care-givers and other residents) [8-11].

In general, personal interest in technology (PIIT) is an important factor for acceptance and usage of technologies [12]: there appears to be a relationship between PIIT and perceived enjoyment, suggesting that the more people are interested in new technologies, the more enjoyment they perceive while using new technologies [12].

In conclusion, several factors appear to play an important role in the acceptance and usage of social robots: (1) Utilitarian factors, such as ease of use and usefulness. (2) Hedonic factors such as enjoyment and playfulness. (3) Social factors such as approaching robots in groups and communicating about robots with family and friends. (4) Personal interest in technology. No studies were found that looked at the same combination of above factors. To fill this gap, this study will look at utilitarian-, hedonic- and social factors and PIIT.

Relationships with robots. Relationships between humans and robots are assumed to be very important predictors of acceptance and use of robots. Many studies investigated relationships with social robots, studying robots such as (1) AIBO, a robot resembling a dog [13], (2) Robovie, a humanlike robot [14], and (3) Paro, a seal robot used for animal assisted therapy with elderly people suffering from dementia, [9, 10]. One of the studies showed that it was possible to build a relationship with AIBO: 70-80 percent of AIBO owners felt a strong attachment to AIBO [11]. Another example of a relationship with AIBO is described in [13], where a girl nurtured an AIBO all the time and saw AIBO as a living being [13]. The same study also indicated that there are two types of relationships humans have with robots: either humans love and nurture social robots and built relationships with them, or humans see social robots as artificial, as a machine [13]. For example, one elderly man treated a robotic doll as if it was his ex-wife and loved and nurtured the robotic doll, while another elderly man saw the robotic doll as an interesting artefact and he slapped it just to see what would happen. Another interesting difference observed was that humans either talked to the robot or about the robot. The elderly man who saw the robotic doll as an artefact talked about the robot when interacting with the researchers, while the elderly man who saw the robotic doll as if it was its ex-wife talked directly to the robot itself [13].

Research with Robovie showed that not every participant was able to build a relationship with the robot, but there were also examples found of children who were indeed able to build a relationship with it [14].

Elderly people were also able to build a relationship with Paro [10] [15]. For example, elderly people stated that they felt better after Paro was introduced in their nursing home. They felt as if they had a new playmate and felt less lonely [15]. Another example: *"Some residents expressed a special attachment to Paro. They spoke to it like it was a pet, gave it names and engaged it in (one-sided) conversations[..] These users generally began a relationship with Paro in which they saw it as dependent of them. Very often they are/were pet owners."* [10, pp. 3]

In conclusion, people's interactions with robots should be studied long-term to establish whether relationships with social robots occur. Indicators from the literature for the presence of a relationship are (1) whether people love and nurture social robots instead of seeing it as artificial and (2) whether people talk to the robot instead of talking about the robot. Until now, most studies with robots studied interaction with social robots in controlled experiments. Only few studies looked at the use of social robots over a long period of time in domestic environments (e.g. [23] [25], where participants were studied in a domestic environment for 3-6 weeks and [29] where participants were studied in a domestic environment for a period of six months.) We believe that observation over a long time period is necessary to study whether people can build (long-term) relationships with social robots [3]. Therefore, this study will look at usage of robots in three different studies with an interaction period of 10 days per study, over a period of one year. The number of participants will grow during time: three during the first study, six during the second study (the three first participants and three new participants), and nine during the third study (the six participants that participated in study 1 and 2 and three new participants). In this paper the first study is reported.

Research questions. The purpose of this study is to gain insight into how people use social robots for health-promotion at home and, in particular, whether people are able to build relationships with contemporary social robots. Consequently, the main research questions of this study are:

(1) "How are social robots used by elderly people in a domestic environment?"
(2) "Which factors play a role in building and maintaining a relationship with social robots?"

3   Methods

Artifacts. The social robot used in this study is the Violet´s Nabaztag, type Nabaztag:tag: a rabbit-shaped Wi-Fi enabled ambient electronic device (www. nabaztag.com). The Nabaztag has no mechanisms of learning or memory. The Nabaztag is able to receive pre-defined spoken commands, but it is not able to understand natural language. However, through its ability to be programmed, the Nabaztag can serve as a robotic user interface to intelligent applications that make use of external sensors and program. Personalized activity plans provided by the participants were used as input for personalized heath related conversations provided by the Nabaztag Participants could respond to these programmed conversations and messages via yes- and no-buttons that were added to the set-up of the Nabaztag.

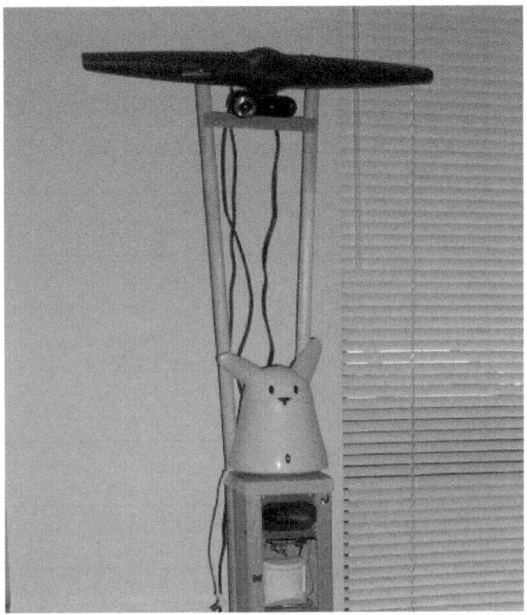

Fig. 1. Set-up of the Nabaztag

Procedures. The Nabaztag was installed for 10 days at the participants' homes. The goal for participants was to improve their overall health condition. Conversations regarding health activity were initiated at four different times of the day and participants also received a daily weather report and messages from the researchers, e.g. *"did you have a good time at the town market?"* All conversations regarding health activities ended with closed questions. Participants could respond to the Nabaztag via the yes- and no-buttons. All conversations between the social robot and the participants were video recorded. Participants received a compensation of £20 for energy costs made during the study.

Participants. An announcement was placed at a website aimed at people older than 50 years old, living in the United Kingdom. Three participants volunteered to take part in this study. The participants (n=3) were all female and between 50-65 years old. The education of the participants differed, from formal education until 16 years old, to a Bachelor degree and a Master degree. One participant was retired, the other participants had a job.

Material. After the 10-day interaction period, experiences of participants were evaluated via a semi-structured interview (Table 1).

Table 1. Used topics/categories during the interviews

Topics	Categories
General use of Nabaztag	Intention to usage [16] Usefulness [16] Usage [16] Expectations Health exercises Evaluation of the possibilities of the Nabaztag (usefulness of design)
Communication with the Nabaztag	Perceived enjoyment [1] [12] Perceived playfulness [17] [18]
Relationship development with the Nabaztag	Trust [19] Likeability [19] Source credibility [19] Appearance (and the uncanny valley) Relationship building Novelty effect
Social factors	Subjective norm [16] Self-identity [16]
Personal interest in technology	Personal interest in technology [12]

Coding and reliability. Linear- and cross-sectional analysis was used to analyze the interview data [20]. After the interviews, the audio recordings of the interviews were transcribed verbatim. The transcriptions were categorized via the used categories of Table 1. After analyzing the interview data, the video data was analyzed [21]. First, the videos were watched. No coding system was designed due to the explorative nature of this study. After watching the videos, the researchers discussed the findings with each other via the visual images. After analyzing the video data, the results of the video data were used to verify/disconfirm the results of the interviews and vice versa.

4   Results

Utilitarian Factors. We refer to participants as A, B, and C. Participants A and B stated in the interviews that they did not find the Nabaztag a useful device because of technical problems and the limited conversation abilities. An indication was found that Participant C found the Nabaztag useful, because she stated in the interviews that she did find the goal of the Nabaztag, for research purposes, useful. All participants found the Nabaztag easy to use.

Hedonic Factors. Participant A and B stated in the interviews that they did not perceive such factors. Participant C stated in the interviews that it was fun to use the rabbit. All participants stated in the interviews that they did not perceive playfulness.

Social Factors. (1) All participants stated in the interviews that they discussed the Nabaztag with others. (2) The videos showed that Participants A and C did not show the Nabaztag to family and friends, but they stated in the interviews that they did show photographs of the Nabaztag to them. The videos showed that Participant B showed the Nabaztag to family and friends. (3) The videos showed that Participants A and C interacted alone with the Nabaztag when interacting with it for the first time and that Participant B interacted with her partner when interacting with the Nabaztag for the first time.

General Usage. The videos showed that Participant C seemed to have embedded the Nabaztag into everyday life, e.g. combining household tasks and interaction. The videos also showed that Participant C experimented with the Nabaztag and found ways to trick the Nabaztag, e.g. by using spare keys when leaving the house for groceries or pushing the no-button when the Nabaztag asked whether participant C had a good time when doing exercises.

Interaction. When communicating with the Nabaztag, the videos showed that Participant A used verbal communication, Participant B used non-verbal communication and Participant C used both verbal- and non-verbal communication. Examples of used non-verbal communication of Participants B and C were mimicking the rabbit and waving to it when leaving the house.

Personal Interest in Technology. The interview data showed that Participants A and B could be categorized as early adopters and that Participant C could be categorized as belonging to the majority regarding adoption of technology [22].

Relationships. Participant C stated in the interviews that she did build a relationship with the Nabaztag: she gave it a name, "Harvey"; she found the rabbit enjoyable to use and interacted with it via both verbal and non-verbal behaviour. The relationship was described in the interview as: *"He asked, the questions, I answered them."* Participant C stated in the interviews that she did not see the Nabaztag as a friend. *"[…] He's a man-made presence or even a women-made presence, in my kitchen […]"*.

5   Discussion

The findings of this study showed that Participants A and B did not perceive a lot of utilitarian factors and that Participant C did show some hedonic factors when using the social robot and was able to built a relationship with the Nabaztag. It is assumed that there is a relationship between showing hedonic factors by participants when using a social robot and being able to build a relationship with these robots.

Regarding the social factors of usage of social robots (1) All participants discussed the Nabaztag with family and friends; (2) Participant B also tended to show the Nabaztag to family and friends, similar to [23] where Roomba, a vacuuming robot, was used to show off. This could imply that Participant B was showing off. Participants A and C showed photographs of the Nabaztag to family and friends. All these results implied that all participants did not see the Nabaztag as a simple piece of technology [24]. (3) Participants A and C interacted individually with the Nabaztag the first time. This finding differs from [6], probably due to the fact that the data of the earlier mentioned studies were gathered in a public area instead of the domestic area. Participant B did not interact alone with the Nabaztag the first time. This could be due to the fact that Participant B did not live alone like the other participants. Furthermore, when looking at Personal Interest in Technology. Participants A and B, early adopters, did not perceive utilitarian and hedonic factors when interacting with the Nabaztag. Participant C, belonging to the majority in adoption of technology, did perceive utilitarian and hedonic factors. This could imply that Participants A and B had higher expectations than Participant C regarding the Nabaztag, and that that expectations were not realized for Participant A and B.

Regarding the usage of social robots in general, Participant C seemed to have embedded the Nabaztag into everyday life. This could be due to the physical place where the Nabaztag was situated in the participant's house. Participant C also experimented with the Nabaztag by tricking it. Similar results were found when studying usage of Roomba in domestic environments [25]. Experimentation with new technology can be an indication for the appropriation of technology [26]. Furthermore, when observing the interaction, Participants used verbal and non-verbal communication when interacting with the Nabaztag, e.g. waving to the Nabaztag when leaving the house. This could imply that human-human communication was used when interacting with the Nabaztag, like was argued in [27] [28], namely that social rules guiding human-human interaction can be applied to human-computer interaction.

With respect to building a relationship with social robots, Participant C stated that she was able to build a relationship with the Nabaztag. This could indicate that relationship building with robots is related to acceptance of the robots. The results showed that relationship building seemed to be related to (1) naming the Nabaztag, also shown in [25], (2) perceiving utilitarian- and hedonic factors and (3) using both verbal and non-verbal behaviour when interacting with the Nabaztag. Participant C showed all these behaviors, while Participants A and B did not show any of these behaviors.

6   Conclusion, Limitations and Future Research

The goal of this study was to get more insight in how elderly people use social robots in domestic settings in general and particularly whether elderly people are able to build relationships with these robots. This study yielded interesting insights such as (1) utilitarian-, hedonic-, and social factors seem important reasons for participants to accept social robots in their domestic environments, (2) the physical location where a social robot is situated in a home is of importance for the ease of acceptance and use, (3) utilitarian-, hedonic- and social factors are important for building a relationship with social robots, (4) there seems to be a relationship between name-calling and relationship building with social robots and (5) there seems to be a relationship between the use of non-verbal and verbal communication of participants and relationship building with social robots.

Since this was only the first, explorative study in a series of studies, the conclusions will be further explored in studies which are currently being undertaken. Our main goals in these studies are: (1) to establish whether utilitarian-, hedonic- and social factors are important in accepting social robots, (2) to explore whether utilitarian-, hedonic and social factors are important for building a relationship with social robots, (3) to explore the relationship between name-calling and relationship-building and (4) to explore the relationship between the usage of non-verbal and verbal communication and relationship-building.

A limitation of this study was that the participants did not find the Nabaztag useful, because it did not help participants to improve their overall health due to technological problems and a limited activity plan. These problems should be solved before the next iteration. Another limitation was the small number of participants. But small, qualitative studies are an essential step to provide in-depth insight into this phenomenon. Although there are still many interesting questions unanswered about which factors are important for the acceptance and usage of social robots, this study did provide a rich, first understanding of how people use social robots in domestic environments.

Acknowledgments. The research leading to these results has received funding from the European Community's Seventh Framework Programme [FP7/2007-2013] under grant agreement no. 231868 and project name Social Engagement with Robots and Agents (SERA).

References

1. Davis, F.D., Bagozzi, R.P., Warshaw, P.R.: Extrinsic and Intrinsic Motivation to Use Computers in the Workplace. Journal of Applied Social Psychology 22(14), 1111–1132 (1992)
2. van der Heijden, H.: User Acceptance of Hedonic Information Systems. MIS Quarterly 28(4), 695–704 (2004)
3. Fong, T., Nourbakhsh, I., Dautenhahn, K.: A Survey of Socially Interactive Robots. In: Robotics and Autonomous Systems, pp. 143–166 (2003)
4. Leite, I., Martinho, C., Pereira, A., Paiva, A.: iCat, an Affective Game Buddy based on Anticipatory Mechanisms (short paper). In: Proc. Of 7th Int. Conf. on Autonomous Agents and Multiagent Systems, pp. 1229–1232 (2008)
5. Looije, R., Neerinckx, M.A., de Lange, V.: Children's Responses and Opinion on 3 Bots that Motivate, Educate and Play. Journal of Physical Agents 2(2), 13–20 (2008)
6. Shiomi, M., Kanda, T., Ishiguro, H., Hagita, N.: Interactive Humanoid Robots for a Science Museum. In: HRI 2006, pp. 305–312 (2006)
7. Weiss, A., Bernhaupt, R., Tscheligi, M., Wolherr, D., Kuhnlenz, K., Buss, M.: A Methodological Variation for Acceptance Evaluation of Human-Robot Interaction in Public Places. In: The 17th IEEE International Symposium on Robot and Human Interactive Communication, pp. 713–718 (2008)
8. Robins, B., Dautenhahn, K., te Boekhorst, R., Billard, A.: Robots as Assistive Technology, Does Appearance Matter? In: Proceedings of the 2004 IEEE International Workshop on Robot and Human Interactive Communication (2004)
9. Shibata, T., Wada, K., Ikeda, Y., Sabanovic, S.: Tabulation and Analysis of Questionnaire Results of Subjective Evaluation of Seal Robot in 7 Countries. In: The 17th IEEE International Symposium on Robot and Human Interactive Communication, pp. 689–694 (2008)

10. Kidd, C.D., Taggart, W., Turkle, S.: A Sociable Robot to Encourage Social Interaction among the Elderly. In: International Conference on Robotics and Automation, pp. 3972–3976 (2006)
11. Fujita, M.: On Activating Human Communications with Pet-type Robot Aibo. Proceedings of the IEEE 92(11) (2004)
12. Serenko, A.: A Model of User Adoption of Interface Agents for Email Notification. Interacting with Computers, 461–472 (2008)
13. Turkle, S., Taggart, W., Kidd, C.D., Dasté, O.: Relational Artefacts with Children and Elders, the Complexities of Cyber Companionship. Communication Sciences 18(4), 347–361 (2006)
14. Kanda, T., Sata, R., Saiwaki, N., Ishiguro, H.: A two-month Field Trial in an Elementary School for Long-term Interaction. IEEE Transactions on Robotics 23(5), 962–971 (2007)
15. Wada, K., Shibata, T.: Robot Therapy in a Care House, Results of Case Studies. In: The 15th IEEE International Symposium on Robot and Human Interactive Communication, September 6-8, pp. 581–586 (2006)
16. Lee, Y., Lee, J., Lee, Z.: Social Influence on Technology Behaviour, Self-identity Theory Perspective. The DATA BASE for Advances in Information Systems 27(2&3), 60–75 (2006)
17. Kim, J.W., Moon, Y.-G.: Extending the Technology Acceptance Model for a World-Wide-Web Context. Information & Management 38, 217–230 (2001)
18. Ahn, T., Ryu, S., Han, L.: The Impact of Web Quality and Playfulness on User Acceptance of Online Retailing. In: Information and Management, pp. 263–275 (2007)
19. Rau, P.L., Li, Y., Li, D.: Effect of communication style and culture on ability to accept recommendations from robots. Computers in Human Behaviour, pp. 587–595 (2009)
20. Mason, J.: Qualitative research, 2nd edn. Sage, London (2002)
21. Jacobs, J.K., Kawanaka, T., Stigler, J.W.: Integrating qualitative and quantitative approaches to the analysis of video data on classroom teaching. International Journal of Educational Research 31(8), 717–724 (1999)
22. Rogers, E.M.: New Product Adoption and Diffusion. Journal of Customer Research 2, 290–301 (March 1995)
23. Forlizzi, J.: ZIB How Robotic Products become Social Products: an Etnographic Study of Robotic Products in the Home. In: ACM/IEEE International Conference on Human-Robot Interaction (2007)
24. Quin, C.: The Emotional Life of Objects. The Journal of Design and Technology Education 8(3), 129–136 (2003)
25. Forlizzi, J., DiSalvo, C.: Service Robots in the Domestic Environment: A Study of the Roomba Vaccuum in the Home. In: HRI 2006, pp. 258–266 (2006)
26. Caroll, J., Howard, S., Vetere, F., Peck, J., Murphy, J.: Just Do What the Youth Want? Technology Approriation by Young People. In: Proceedings of the 35th Hawaii International Conference on System Sciences (2002)
27. Nass, C., Moon, Y., Fogg, B.J., Reeves, B., Dryer, C.: Can Computer Personalities be Human Personalities. In: CHI 1995 Mosaic of Creativity, pp. 228–229 (1995)
28. Lee, M.K., Kiesler, S., Forlizzi, J.: Receptionist or Information Kiosk: How do People Talk with a Robot? In: The 2010 ACM Conference on Computer Supported Cooperative Work, pp. 31–40 (2010)
29. Sung, J.-Y., Grinter, R.E., Christensen, H.I.: Pimp my Roomba. Designing for Personalization. In: CHI 2009, Boston, MA, USA (2009)

The Potential of Socially Assistive Robotics in Care for Elderly, a Systematic Review

Roger Bemelmans[1], Gert Jan Gelderblom[1], Pieter Jonker[2],
and Luc de Witte[1,3]

[1] Zuyd University, Nieuw Eyckholt 300, 6419DJ Heerlen, The Netherlands
[2] TU Delft, Mekelweg 2, 2628CD Delft, The Netherlands
[3] Maastricht University, Universiteitssingel 40, 6229ER Maastricht, The Netherlands
r.bemelmans@hszuyd.nl, g.j.gelderblom@hszuyd.nl, p.p.jonker@tudelft.nl,
l.de.witte@hszuyd.nl

Abstract. The ongoing development of robotics against the background of a decreasing number of care personnel raises the question which contribution robotics could have to rationalize and maintain, or even improve the quality of care.

A systematic review was conducted to assess the effects and consequences of the interaction between socially assistive robots and elderly, in published literature. We searched in CINAHL, MEDLINE, The Cochrane Library, BIOMED, PUBMED, PsycINFO, EMBASE and the IEEE Digital Library. In addition, articles were selected through free Internet search and from conference proceedings.

Studies have been found reporting positive effects of companion type robots on (socio)psychological level and physiological level. However, the scientific value of the evidence is limited due to the fact that most research is done in Japan with a small set of robots, with small sample sets and with mostly an explorative approach.

Keywords: Robotics, effects, effectiveness, elderly, interventions, literature review.

1 Introduction

The ongoing development of technology, specifically robots, against the background of a decreasing number of care personnel raises the question what the potential contribution of robotics could be in rationalizing and maintaining, or even improving the quality of care. Robots can contribute to health care support in terms of capacity, quality (performing very accurately and task specific), finance (support or even take over tasks of trained personnel) and experience (e.g. increase feeling of autonomy and self management).

The idea of robotics playing a role in health care was launched some decades ago and has mainly been developed for physical training in rehabilitation as well as personal assistance for ADL tasks [2]. Robotic applications supporting social behavior are a more recent development [1]. Marti et al. [5] describe these

M.H. Lamers and F.J. Verbeek (Eds.): HRPR 2010, LNICST 59, pp. 83–89, 2011.

socially assistive robots (SAR) as being capable of mediating social interaction, not designed to help the human being performing work tasks or saving time in routine activities, but to engage people in personal experiences stimulated by the physical, emotional and behavioral affordances of the robot. So far systems have been developed supporting child's play (e.g. [4]) and care for elderly with dementia (e.g. [6]).

When applying the ICF-classification [7] socially assistive robots are the Environmental Factors (e) in the context of Activities and Participation (d). The domains we are interested in are General Tasks and Demands (d2), Communication (d3), Interpersonal Interactions and Relationships (d7) and Recreation and Leisure (d92).

To reach a better matching between robot technology and the needs of elderly care, we performed a study to obtain insight in the potential of socially assistive robotics for elderly care.

2   Search Method

In September 2009 the CINAHL, MEDLINE, Cochrane, BIOMED, PUBMED, PsycINFO and EMBASE databases and the IEEE Digital Library (Xplore) were explored. No limitations were applied for date of publication. Only papers written in English were taken into account.

Selected articles went through a selection process, based on title, abstract and complete content, in order to obtain a final set of articles to be included in the review. The objective of the search, in short, was to find measured effects and consequences of socially assistive robots used in interventions in elderly care. The search query was divided into three logical conjunctive components. These components represent, with several free words and Medical Subject Headings (MeSH) terms, the objective (measured effects), the subject (elderly) and the means (robots).

To limit the chance of excluding relevant articles the search in the first step was based solely on subject and means, so the objective (measured effects) was not included. The free words for the subject (or their database specific Thesaurus equivalent) were "elder*", "age*", "old people", "senior*" and "dementia" and their associated MeSH terms (or their database specific equivalent) were "Housing for the Elderly", "Aged", "Health Services for the Aged", "Residential Facilities" and "Dementia" (including their subheadings). The free words for the means (or their database specific Thesaurus equivalent) were "robot*" and "assis* technol*" and their associated MeSH terms (or their database specific equivalent) were "Robotics", "Self-Help Devices" and "Mobile Health Units" (including their subheadings). By using the asterisk (*) the term becomes a prefix. So 'assis*' represents among others 'assisting' and 'assistive'. In a second step three reviewers individually selected relevant articles, based on their title for the third selection. In a third step the articles were individually judged by the three reviewers based on their abstracts. In a fourth step the articles were read in full and judged by one reviewer on order to obtain the final set of articles for the review.

In addition articles were selected through free Internet search (Google, Google scholar), and by hand from conference proceedings (HRI, ICORR, ICRA, ROMAN) and reference lists of selected articles.

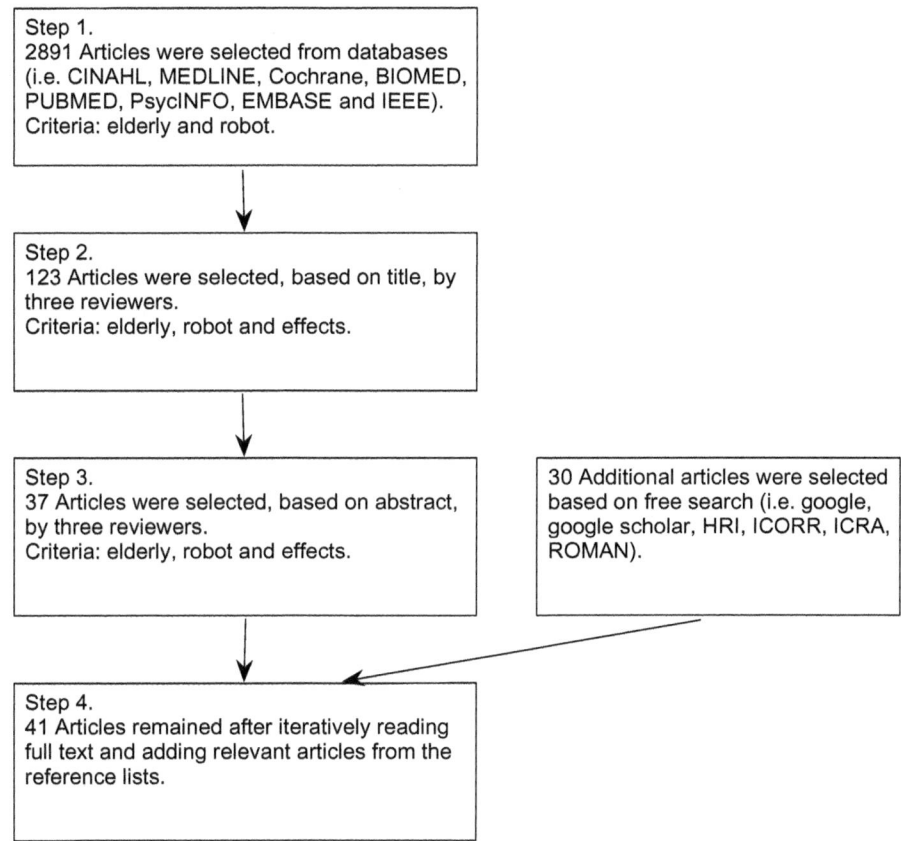

Fig. 1. Schematic overview of selection process with search results

3   Results

In the first step 2891 articles were found in the aforementioned databases. In the second step three reviewers individually selected 123 relevant articles for the third step. In the third step 37 articles were selected, based on their abstracts. In addition, 30 articles were selected via free Internet search and from conference proceedings. Finally 41 articles, of which 30 from step three, were included in the review. See figure 1.

The 41 included articles contain 16 studies and involve 4 robot systems, and 1 undefined robot. There are 9 journal articles, 2 electronic articles and 30 conference proceedings. Categorizing the articles based on the robot system there are 3 articles on the robot Bandit describing 1 study (by Tapus et al.), 5 articles on the Aibo robot describing 5 studies and 29 articles on the Paro robot describing 7 studies (majority by Wada, Shibata et al.). Furthermore 2 articles describing 2 studies about the robot NeCoRo (all by Libin et al.), 1 article with an unspecified robot and 1 article with an overview of several robots were selected. Table 1 presents an overview of the characteristics of the aforementioned robots.

In the following paragraphs the main studies per robot system are shortly described.

Paro. The majority of the selected articles, involving the seal robot Paro, describe two studies. In the first study the seal robot was given to 14 elderly in a health service facility. A desk was prepared for the robots in the center of a table. They interacted freely with the robot for about 1 hour per day and 2 days per week, over a period of 1 year. The results showed that interaction with Paro improved their moods and depression, encouraged their communication, decreased their stress level, and then the effects showed up through one year. In the second study the experiment has been conducted in a care house, 12 participated aged from 67 to 89 years. Caregivers activated Paro on a table in a public space, each day at 8:30 and returned to their office until 18:00, for a period of 2 months. The residents could play with Paro whenever they wished during the time. The results show that Paro encouraged them to communicate with each other, strengthened their social ties, and brought them psychological improvements. Physiologically, urinary tests showed that the reactions of their vital organs to stress were improved by the introduction of Paro.

NeCoRo. One pilot study compared the benefits of the robotic cat and a plush toy cat as interventions for elderly persons with dementia. The study consisted of two interactive sessions, one with the robotic cat and one with the plush cat, with a duration of 10 minutes each. Only one session per day was conducted for each of the 9 participants. The sessions were presented in random order in an attempt to rule out the potential novelty effects. Increases of pleasure was measured. It also showed that the cats hold promise as an intervention for agitated behaviors. The amount of physically disruptive behaviors and overall agitation decreased significantly when residents interacted with the cats. This study was limited by its small sample size and short-term sessions. Another study was cross-cultural oriented, regarding American and Japanese perceptions of and communications with the robotic cat. The participants, 16 Americans and 16 Japanese of both genders and two age groups (20−35 and 65−79), interacted individually with the robot in a 15 minute session. It seems that Americans enjoy touching the robotic cat a little bit more than the Japanese. Males from both cultures, more so than females, like the cats active behavior. Past experience with real pets was positively associated with the interest in the robotic cat. This study was limited by its short-term sessions.

Table 1. Socially Assistive Robots used in studies

robot	description	fig
NeCoRo	A cat-like robot with synthetic fur, introduces communication in the form of playful, natural exchanges like between a person and a cat. Via internal sensors of touch, sound, sight and orientation human actions and its environment can be perceived. Behavior is generated, based on internal feelings, using 15 actuators inside the body.	
Bandit	A humanoid torso mounted on a mobile platform. The mobile platform is equipped with a speaker, color camera and an eye-safe laser range finder. The torso includes: two 6 Degrees Of Freedom (DOF) arms, two 1 DOF gripping hands, one 2 DOF pan/tilt neck, one 2 DOF pan/tilt waist, one 1 DOF expressive eyebrows and a 3 DOF expressive mouth. All actuators are servos allowing for gradual control of the physical and facial expressions.	
AIBO	A dog-like robot that can see, hear and understand commands. It has the ability to learn, to adapt to its environment and to express emotion. It uses its Illume-Face to communicate when it detects toys, someone's hand, voice commands or face and voice. Each expression appears as an animated pattern on the Illume-Face display, created by LEDs that light up or fade out to varying degrees.	
Paro	A seal-like robot with five types of sensors: tactile, light, audio, temperature and posture, with which it can perceive people and its environment. With the light sensor it can distinguish between light and dark. It feels being stroked or beaten by its tactile sensors, or being held by the posture sensor. It can recognize the direction of voice and words such as its name and greetings with its audio sensor.	

Bandit. The reported study focuses on the possible role of a socially interactive robot as a tool for monitoring and encouraging cognitive activities, in comparison to a computer screen, of elderly suffering from dementia. The social therapist robot tries to provide customized cognitive stimulation by playing a music game with the user. The study consisted of a 20 minute session per week for 8 months, with 3 participants. Each session involved supervised instructed music based cognitive games. Improvement was observed for all participants with respect to reaction time and incorrectness, proportional with their level of cognitive impairment. The participants enjoyed interacting with the robot and preferred the robot to the computer screen. Also the ability of the participants to multitask (singing and pushing button at the same time) was reported. The results are not conclusive because of the small number of participants used in the study.

AIBO. Several studies about the use of AIBO within elderly care have been carried out, including studies in which the robot was compared to toy and living dogs. The results indicate that robot-assisted activity is useful to reduce loneliness and improve activities and emotional state of elderly people with dementia. On the other hand, the absence of a soft skin and the limited response capability to touch stimuli is also reported.

4   Conclusions

The reported literature review produced a limited set of studies for which a wide search was required. The domain of socially assistive robotics and in particular the study of their effects in elderly care has not been studied comprehensively and only very few academic publications were found. The studies that were found were mainly reported in proceedings underlining the initializing stage of the application of this type of robot system.

In the reported studies a small set of robot systems were found to be used in elderly care. Although individual positive effects are reported, the scientific value of the evidence is limited due to the fact that most research is done in Japan with a small set of robots (mostly Paro and AIBO), with small sample sets, not yet clearly embedded in a care need driven intervention. The studies were mainly of an exploratory nature, underlining once more the initial stage of application within care.

In general relations between the type of outcomes aimed for, either related to support of care or support of independence and the application of the robot system in care are not well established. Care interventions are adopted within health care systems because of their added value to the provision of care. The reported outcomes only partly were directly linked to desired outcomes, materializing the desired added value.

Nonetheless, the potential of the robot systems seems generally accepted, based on the initial results and face value. There seems to be potential for added value. To establish these, additional research is required experimentally investigating the effects of interventions featuring socially assistive robots within real elderly care setting.

References

1. Butter et al.: Robotics for health care, final report. Technical report, TNO, report made for the European Commission (2008)
2. Butter et al.: Robotics for health, state of the art report. Technical report, TNO, report made for the European Commission (2007)
3. Higgins, J.P.T., Green, S. Cochrane handbook for systematic reviews of interventions (2008)
4. Interactive Robotic social Mediators as Companions (Iromec) project (2009), http://www.iromec.eu
5. Marti, P., Bacigalupo, M., Giusti, L., Mennecozzi, C., Shibata, T.: Socially Assistive Robotics in the Treatment of Behavioural and Psychological Symptoms of Dementia. Paper presented at the Biomedical Robotics and Biomechatronics (2006)
6. Wada, K., Shibata, T., Asada, T., Musha, T.: Robot therapy for prevention of dementia at home. Journal of Robotics and Mechatronics, 691–697 (2007)
7. World Health Organization.: International Classification of Functioning, Disability and Health (ICF). WHO Press (2001)

The Yume Project: Artists and Androids

Michael Honeck, Yan Lin, David Teot, Ping Li, and Christine M. Barnes

Carnegie Mellon University, Entertainment Technology Center
700 Technology Drive, Pittsburgh PA, 15219
{mhoneck,yanlin,dteot,pingl,cbarnes1}@andrew.cmu.edu

Abstract. Creating believable androids is not just a technological feat, but an artistic one. The Yume Project addresses the challenge of creating lifelike androids by developing complex character. The authors demonstrate the unique perspectives artists can bring to the humanoid robot design process through costume, character, and story development. The authors hope to show that it is a focus on believability rather than realism that is needed to advance the field of humanoid robotics.

Keywords: humanoid robotics, interdisciplinary teams, animatronics, art, human-robot interaction, entertainment technology, android.

1   Introduction

The Yume Project takes a unique approach to the challenge of creating androids that appear human in both appearance and interaction. We are not designing new systems, but rather seeing how existing android technology can be enhanced by artistic disciplines [1]. This strategy is worth examining because it brings a new perspective to a field that can prosper from greater interdisciplinary teamwork [2,3,4]. We seek to show that android design would benefit from a shift in focus from realism to believability. We propose that to engage humans, a compelling, believable character is as important as an accurate simulation of human physiology and intelligence [5]. Preliminary data will also be presented that suggests that the enhancements made resulted in audience members finding the robot more appealing on initial viewing.

1.1   Project Scope

The Yume Project was a semester-long endeavor. The team was composed of artists specializing in theater and film production, a writer, and a computer programmer. This mix is unusually art-heavy for a robotics project. This provided the team unique perspective of the issues encountered when working with androids. The hardware used on the project was Kokoro Robotics' DER-1 Actroid. The Actroid is a pneumatically driven female android covered with a partial silicone skin. Throughout the course of the semester, the team focused on creating inexpensive proofs-of-concept that could be rapidly implemented and iterated upon.

M.H. Lamers and F.J. Verbeek (Eds.): HRPR 2010, LNICST 59, pp. 90–97, 2011.

2   Believability versus Realism

When making attempts at verisimilitude, slight inaccuracies become large distractions as a more realistic form creates ever higher expectations about mobility and intelligence [3,4]. Total realism then, would only be achieved in the completely accurate recreation of the human form. Our knowledge of the human form, especially the mind, is not complete. Coupled with technological limitations, meeting expectations with realism becomes a complex and tedious task.

Believability however, comes in the successful communication of the human condition. Consider memorable characters created in novels, theater, and film. They seem full of life; they meet our expectations, because we identify with their challenges and struggles [6]. For the creators of these characters, realism becomes a tool in service of believability [7].

By making believability, rather than realism, the ultimate goal of android development, we are not abandoning the quest for ever more perfect technology, we are simply attempting to meet human expectations of androids with a new set of tools. In the Yume Project, technology was held constant. This provided us the opportunity to focus on character and story in order to increase the believability of an android.

3   Building Character

An understanding of character is the starting point for believability. Thinking about character: a person's goals, quirks, and background [8], can help inform the required functionality, appearance and personality of an android. These character goals can also be designed to meet research goals. Because the Yume Project is working with pre-existing hardware, a character was developed that also embraced our system's limitations.

3.1   Character

Humans naturally perceive some degree of personality in robots [9]. By developing character for androids, we are embracing this tendency. To develop our Actroid's personality we drew from our own experiences, personalities, and objectives [10]. We named the robot Yume (pronounced you-meh or you-me), the Japanese word for "dream."

Using a character sheet [8], we developed a background for Yume. The character sheet serves as a history to reference in order to keep voice, personality and appearance consistent across interactions. In Yume's character sheet we describe her appearance, occupation, and personality in detail, including grooming habits, favorite band and idiosyncrasies. Not every facet of character needs to be revealed to a person interacting with the robot. The information has value as subtext, providing reason for a character's actions [11]. Without this subtext, any response generated by an android has the potential to seem disingenuous and unbelievable [12].

One advantage of exploring character in this manner is that technical limitations can be incorporated into personality. For example, the Actroid is pneumatic which can cause jerking movements, and does not always make eye contact with viewers.

We looked for humans that act similarly and found anecdotally that members of the Goth subculture dance with jerking movements and avoid making eye-contact. We felt that making Yume a member of this subculture would turn these limitations into character traits that make sense to viewers.

3.2 Story

Character will reveal itself through an android's interactions [6]. Reactions that are consistent with character satisfy expectations and create believability. These traits could certainly be incorporated into AI systems, but believability can also be enhanced by simplifying interaction. As an example, the Actroid is equipped with voice recognition software which we chose to disable. Currently, voice recognition technology is not as accurate as natural human conversation. By removing this potentially awkward interaction, a technical limitation disappears and a space is created for the viewer to perceive interaction in carefully designed scripts.

The Yume project used Rogerian therapy techniques similar to those used in the ELIZA program [13] to develop scripts that create the illusion of interaction and intelligence. These techniques simulate interaction by asking rhetorical questions and asking the audience to perform actions. Pauses in dialogue are added to give the impression of thought. These pauses also allow viewers to attribute complex reasoning and emotion to the robot's ambiguous actions [3].

Like our choice of a Goth character, story decisions were made that downplay the Actroid's technical limitations. Because the robot is elevated on a base, we wrote stories that center around addressing a small crowd. Because she is addressing a crowd she should not be making constant eye contact, or interacting heavily with any one person.

4 Bringing Character to Life

In the Actroid, motion is generated through the playback and blending of animations. The team investigated three techniques for creating these animations: face detection feedback loop, motion capture, and traditional character animation.

4.1 Face Detection

When interacting with the Actroid, lack of eye contact was often noted as distracting. To eliminate this distraction, techniques were borrowed from animation and puppetry. A webcam was used to pick faces randomly out of a captured image, then using the animation principle of anticipation [14], the Actroid procedurally turns its eyes followed by its head and then shoulders, to the chosen face. To create the illusion that a character is focusing its gaze ("convergence" in human vision), puppets are built with slightly crossed eyes. This same technique was implemented in the Actroid. In action, motions generated by this feedback loop appeared robotic and extreme.

4.2 Motion Capture

Motion capture has been proposed as a means of duplicating human motion in androids because of its widespread use in film and video games. Attempts to implement such systems have met with unsatisfactory results [15]. Reasons cited for this poor performance include the difference between human and android joint structures, and limited speed and range of joint motion [16]. We propose a more basic reason for these unbelievable performances. In the film and video game industries, raw motion capture data is never translated into final animations. Instead, animators devote hundreds of hours to enhancing motion captured performances. Artists take a completely realistic recreation of human motion and must exaggerate, tweak and accent the performance before audiences find the motion believable and acceptable. Likewise, the completely realistic data generated by motion capture is not the key to believable motion in humanoid robotics without some amount of artistic interpretation.

4.3 Character Animation

Because believable motion is such a large part of conveying understandable character [17], the team decided to put motion creation in the hands of professionals. Animators are motion specialists: expert actors who use the twelve principles of animation originally developed by Walt Disney Studios [14] to convey believable action and character with subtlety.

The Actroid's original animation interface is an internally developed system that does not incorporate the conventions of modern animation software. In order to allow animators to easily apply their skills to android animation, we needed to develop a tool using the vocabulary and conventions of their trade. To meet their needs, we have developed a custom animation rig using Maya (see Fig. 1). Maya is an industry standard animation tool that contains the features animators need to create engaging, believable motion. Maya has been suggested as an animation interface by Hanson robotics [4] and is currently in use for animatronics animation at the Walt Disney Company.

Using the rig we developed, animators now have a three dimensional view of the Actroid that can be directly manipulated. Maya also contains preexisting tools that speed up the content creation process such as the ability to synch sound files with

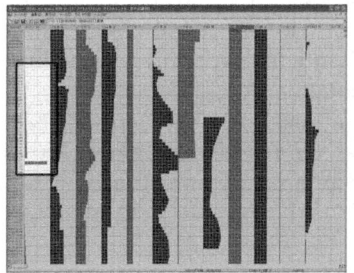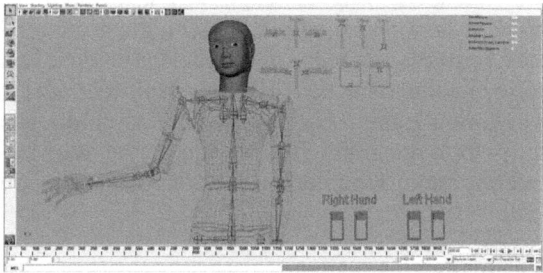

Fig. 1. In the original interface (*at left*), colored curves represent air pressure in actuators. One curve (*highlighted*) represents an eye blink. In our rig (*at right*), a scale model of the Actroid can be animated through direct manipulation. The model simulates the robot's outer skin, range of motion, and internal joint structure. Animation with sound can be previewed at anytime. Facial and hand controls are displayed beside the model to allow animators to more easily see the results of their work.

animation, the ability to interpolate between two keyed poses, and the ability to save libraries of movements that can be blended to create complex actions. Animations can be previewed without the robot present, although the imprecision of the physical world requires animators to tweak their work to play as intended on the Actroid.

5 Presenting Character

Costume is a visible expression of character, it serves as an indicator of social status, style, occupation and age [18]. By carefully designing what our character wears, we can further influence an audience's perception of the Actroid. When this perception is supported by character and story, audience expectations are satisfied, further contributing to believability. For the Actroid, costuming also became an opportunity to downplay distracting physical features (see Fig. 2).

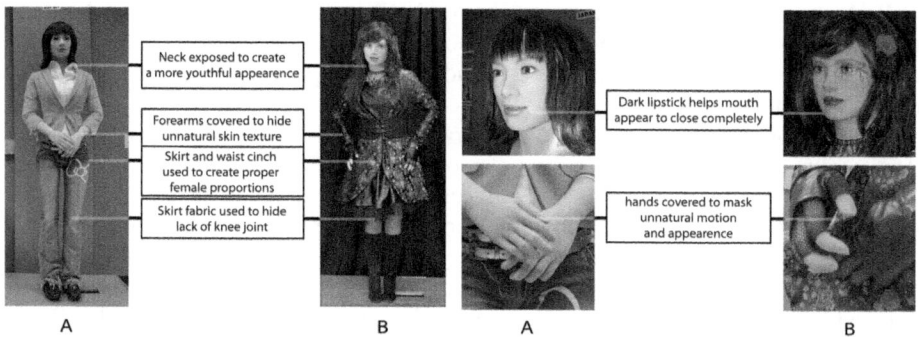

Fig. 2. The Actroid's original clothing (*on left and right center*) and our final costume design (*on right and left center*). Notice the pattern and shine of the final costume's fabric. This helped draw audience attention to the Actroid.

When animated, the Actroid's hands are its least believable feature. In Fig. 2, it is shown that covering the hands with lace and long leather gloves lowers their contrast, making them less likely to attract audience attention. Several casual observers found the Actroid's inability to completely close her mouth unsettling. Dark lipstick corrected this issue by darkening shadows in the mouth, making it less likely a viewer would notice the robot's teeth. The Actroid has no joint at the knee, making weight shifts look unnatural. The design of the skirt hides the robot's upper legs and hips, creating more feminine proportions while moving in a natural way during a weight shift.

5.1 Setting

Setting and lighting can deeply affect a subject's reaction to an android. In film, location and light are used to set a tone and convey emotion. Starlets are lit with warm, soft lights that hide skin flaws and flatter the figure. Taking the same care in presenting androids can be an inexpensive and effective way of preparing an audience for the interaction it is about to have.

Androids are most often encountered in laboratory, academic, and other research environments (Hanson Robotics' PKD android [4] is a notable exception). These settings remind viewers that they are looking at technology. By presenting the Actroid in a manner similar to theater, we hope to evoke the experience of seeing a character in performance (see Fig. 3).

The Actroid is presented surrounded by a semicircle of curtains and bathed in warm lights. Once again, we are attempting to overcome the Actroid's technical limitations. Gold colored lights make the robot's silicon skin look warm and focus audience attention by accenting the costume's gold pattern. Curtains both hide the robot's hardware and suggest a theater environment. When attending a performance, audiences expect to be told a story with little interaction from actors. Recreating this environment prepares the audience for the Actroid's character and limited interaction.

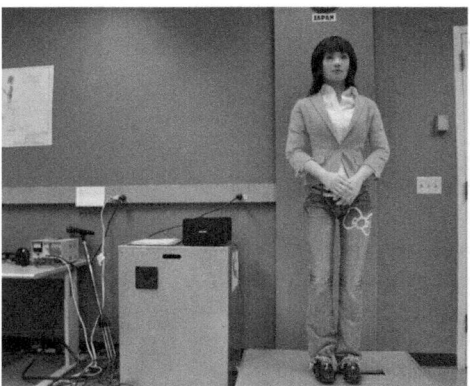

Fig. 3. If the audience feels that they are attending a show instead of witnessing an experiment, they should be less critical of the eccentricities of interacting with an android

6   Audience Response

Over the course of the project, audience members were asked to record their initial reaction to the animated Actroid. Each costume was viewed by a different group of 16 participants. This sample size is too small to draw formal conclusions, but did provide preliminary data which guided the design process.

Reactions were recorded along a 7-point Likert scale (disturbing - unappealing - slightly unappealing – neutral - slightly appealing – appealing - engaging). Upon examining our data, we decided that "disturbing" and "engaging" were misleading word choices, so those responses were treated as "unappealing" and "appealing" respectively, creating a 5 point scale. This did not change results of the analysis.

As can be seen in Figure 4, the final costume elicited an improved response over the initial costume design, $t(15) = 7.66$, $p < .001$. Audience attitudes moved from under "slightly unappealing" to over "slightly appealing," which suggests that our attempts to increase believability by focusing on character and costume had a positive impact on audience perception of the robot.

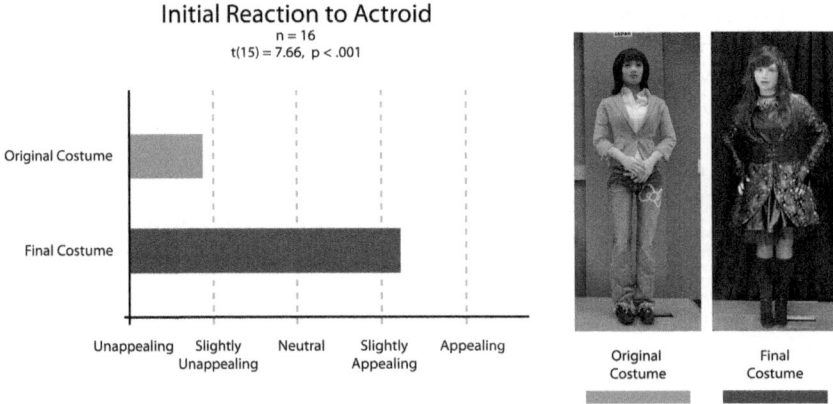

Fig. 4. Participant reactions to different costume iterations, rated from Disturbing to Engaging

7 Conclusion

A perspective shift is needed in android design. Scientists striving for realism are slowed by technology's limits in duplicating the minutia of the human form. By embracing the challenge of creating believability, we feel we can endow androids with strong character that engages audiences while deemphasizing physical and AI weaknesses that would otherwise distract during encounters.

We want researchers to see that artists can have an integral role to play in overcoming the challenge of creating believable humanoid robots. It is not a task engineers and programmers should have to tackle alone. The development of believable androids like Yume requires more than just technical knowledge. Aesthetics, character, and presentation are all equally essential parts of a satisfying human-robot interaction.

References

1. Reichardt, J.: Robots: Fact, Fiction + Prediction, p. 56. Thames & Hudson Ltd., London (1978)
2. Burke, J.L., Murphy, R.R., Rogers, E., Lumelsky, V.J., Scholtz, J.: Final report for the DARPA/NSF interdisciplinary study on human-robot interaction. IEEE Transactions on Systems, Man, and Cybernetics: Part C - Applications and Reviews 34(2), 103–112 (2004)
3. Mondada, F., Legon, S.: Interactions between Art and Mobile Robotic System Engineering. In: Gomi, T. (ed.) ER-EvoRob 2001. LNCS, vol. 2217, pp. 121–137. Springer, Heidelberg (2001)
4. Hanson, D., Olney, A., Pereira, I.A., Zielke, M.: Upending the Uncanny Valley. In: Proceedings of the American Association for Artificial Intelligence (AAII) Conference, Pittsburgh, PA, USA (2005)
5. van Breemen, A.J.N.: iCat: Experimenting with Animabotics. In: Proceedings, AISB 2005 Creative Robotics Symposium (2005)
6. McKee, R.: Story: Substance, Structure, Style, and the Principles of Screenwriting. ReganBooks, New York (1997)

7. Bates, J.: The Role of emotion in Believable Agents. Communications of the ACM 37(7), 122–125 (1994)
8. Harger, B.: Character sheet (unpublished)
9. Turkle, S.: Relational Artifacts with Children and Elders: The Complexities of CyberCompanions. Connection Science 18(4), 347–361 (2006)
10. Stanislavski, C.: Building a Character. Theatre Arts Books, New York (1994)
11. Izzo, G.: Acting Interactive Thater: a Handbook. Heinemann, New Hampshire (1998)
12. MacDorman, K.F., Ishiguro, H.: Toward social mechanisms of android science. Interaction Studies 7(2), 289–296 (2006)
13. Weizenbaum, J.: ELIZA—A Computer Program For the Study of Natural Language Communication Between Man And Machine. Communications of the ACM 9(1), 36–45 (1966)
14. Thomas, F., Johnson, O.: The Illusion of Life-Walt Disney Animation. Walt Disney productions, New York (1981)
15. Ude, A., Man, C., Riley, M., Atkeson, C.G.: Automatic Generation of Kinematic Models for the Conversion of Human Motion Capture Data into Humanoid Robot Motion. In: Proceedings of the IEEE-RAS International Conference on Humanoid Robots, Cambridge, MA, USA (2000)
16. Matsui, D., Minato, T., MacDorman, K.F., Ishiguro, H.: Generating Natural Motion in an Android by Mapping Human Motion. In: Proceedings of the IEEE/RSJ International Conference on Intelligent Robots and Systems, pp. 1089–1096 (2005)
17. van Breemen, A.J.N.: Bringing Robots to Life: Applying Principles of Animation to Robots. In: CHI 2004 Workshop Shaping Human-Robot Interaction, Vienna, Italy (2004)
18. Cunningham, R.: The Magic Garment: principles of costume design. Waveland Press, Illinois (1994)

Digital Adultery, "Meta-Anon Widows," Real-World Divorce, and the Need for a Virtual Sexual Ethic

William David Spencer

Gordon-Conwell Theological Seminary,
130 Essex Street, South Hamilton, Massachusetts, 01982, USA
wspencer@gordonconwell.edu

Abstract. Ethical issues that have emerged around relationships in virtual worlds can inform the way we approach the ethics of human/robot relationships. A workable ethic would be one that treats marriage as an enduring human institution and, while we value robots as worthy works of our hands, they are inappropriate partners for marital or sexual relationships.

Keywords: roboethics.

1 Introduction

If we could only invent some kind of digital device that we could transport across the theorized folds of time, say, into the next century to sample or record the effects of what we are striving to create now, that would help us decide if what we are envisioning and working so hard to bring about will actually be life-enhancing or the opposite.

But, the fact of the matter is that we cannot currently do that directly. Indirectly, however, we can seek out simulations, what we might call meta-tests, standing in analogical parallel to some of the goals we strive to accomplish in robotics. And this is what my paper is about: examining such an analogical simulation, to assess the effect of an experiment already in existence in order to speculate upon the probable effects in the real world of the future of what some are now working to accomplish: robot/human sexual relationships.

My simulated test is the effect on human relationships of digital sexual participation in Second Life. Virtual avatars are my analogy to sexbots. My argument is this: Since none of us can tell for certain what will happen in 2050 and beyond, I ask: Do any comparative parallels currently exist that are not merely speculative but demonstrable? Virtual relationships through avatars seem to me to provide an instructive analogy to sexbots in that they introduce artificially created relational partners into people's lives and, therefore, since we are social creatures, they are often introduced into the networked relationships of families. In numerous documented cases, such liaisons do cause pain for a partner, producing feelings of betrayal and rejection through a loss of attention that affects one's sense of being valued and having one's love and caretaking mutually reciprocated. Here is an overview of the information that leads me to this conclusion.

M.H. Lamers and F.J. Verbeek (Eds.): HRPR 2010, LNICST 59, pp. 98–107, 2011.
© Institute for Computer Sciences, Social Informatics and Telecommunications Engineering 2011

2 Relationships in Virtual Worlds

Philip Rosedale and Linden Lab's Second Life is the most successful of the virtual metaverses inspired by Neal Stephenson's 1992 novel *Snow Crash,* "the novel that taught us to dream about an online digital world that exists in parallel with the corporeal realm," according to James Wagner Au [1, 2006].[1] Millions of participants are now involved, including members of this conference.

Of central interest to everyone participating is the social networking that is its core appeal. The animated replica one creates for oneself is most often the way that one wishes one could be. As Neal Stephenson explained in *Snow Crash,* "Your avatar can look any way you want it to, up to the limitations of your equipment. If you're ugly, you can make your avatar beautiful. If you've just gotten out of bed, your avatar can still be wearing beautiful clothes and professionally applied makeup" [2].

Interaction hyperspaced with the acquisition of the Swedish company "Enemy Unknown" and its "Avatars United," a social network to bring "together" users of "multiple worlds and games" [3]. As avatars interacted and formed relationships, the trajectory paralleled that of real life so that these virtual representations began to date, couple up, marry, and have virtual sexual relationships, not necessarily in that order. All of this, of course, was taking place in a virtual, digital, world, but in many cases the impact was being felt in the real one.

2.1 Impact on Real-World Relationships

Healthy Place is the Web site of the clinical psychologist who founded the Center for Internet Addiction Recovery, Dr. Kimberly Young. It features a set of resources and online communities for sufferers of a variety of psychological maladies from Alzheimer's to self-injury, but has recently added to its resources on cyberspace and Internet addiction books with titles like *Infidelity Online: An Effective Guide to Rebuild your Relationship after a Cyberaffair.* Dr. Young explained to *The Wall Street Journal* that most of the cases she counsels "involve interactive fantasy role-playing games. 'They start forming attachments to other players....They start shutting out their primary relationships'" [4].

What such psychologists and psychiatrists are treating and the courts are now having to litigate are cases like the well-publicized disintegration of the Hoogestraat marriage. While recovering from an illness, the 53-three-year-old husband began spending up to twenty hours a day in the virtual world. Drawing from a background in computer graphics, he accrued several virtual businesses, a staff of twenty-five other people's avatars, a fortune in virtual money, and a virtual wife, the avatar of a 38-year-old Canadian woman. When his real wife of seven months discovered his virtual wife, she was heartbroken and enrolled in EverQuest Widows, an online support group, serving as a kind of "Meta-anon," paralleling Al-Anon for abandoned families. Chronicling the story for the August 10, 2007's *The Wall Street Journal,*

---

[1] Recently, Linden Labs commissioned Fisik Baskerville to design a virtual monument to the novel and its author and place 200 of these in various locations in the landscape. Many of us think immediately of the Matrix movies, while others of such predecessors as Lucasfilm's Habitat or the earlier parallel worlds of imaginative fiction back to the realms of the gods in mythology.

Alexandra Alter reported, "[United States] Family-law experts and marital counselors say they're seeing a growing number of marriages dissolve over virtual infidelity. Cyber affairs don't legally count as adultery unless they cross over into the real world, but they may be cited as grounds for divorce and could be a factor in determining alimony and child custody in some states" [5].

That is actually what did happen the very next year when British newspapers chronicled the divorce proceedings of Amy Taylor and Dave Pollard, two disabled Brits who had met and married virtually on Second Life and then again in real life. Ms. Taylor sued successfully for divorce "under the basis of 'unreasonable behavior,'" because of her husband's continual digital sexual liaisons, after "her lawyer told her that other marriages have also ended over 'Second Life' adultery" [6].

This raises the question: Should a virtual sexual relationship between avatars really be considered adultery if conducted by players with other spouses? Some answer in the negative, as did Yurie, a young Japanese wife, whose husband, Koh, fell in love with a virtual girlfriend from Nintendo's Love Plus dating simulator game. Yurie dismissed "Koh's virtual indiscretions" with, "[I]f he's just enjoying it as a game, that's fine with me" [7]. Here Yurie partially addresses a set of questions posed by Cornell University's Josh Pothen, who asked, "So, does flirting with or engaging with another virtual character really count as adultery, or is it only part of a game? Would both characters have to be controlled by people for it to be adultery? What if one is controlled by a human and the other is computer-controlled? Does it make a difference that the game is an interactive world instead of a regular video game with levels and an ending?" [6]. A dating-game character is not an avatar of another human, but indeed computer-controlled (and here we might draw the analogy to a sexbot).

However, real-world spouses such as Yurie may reconsider such an easy dismissal in light of the well-publicized case of the 27-year-old Tokyo man who did indeed in real life marry a character from the same Love Plus game before "a priest, an MC, a DJ...friends and family" with "photo slideshows, wedding music and even a bouquet" [7]. As "the first human-to-avatar union," the human, who goes by his online name Sal 9000, is very serious about the relationship. He told reporters, "I love this character, not a machine....I understand 100 percent that this is a game. I understand very well that I cannot marry her physically or legally," yet he states he has chosen the character as "better than a human girlfriend." [8]. "Some people have expressed doubts about my actions, but at the end of the day, this is really just about us as husband and wife. As long as the two of us can go on to create a happy household, I'm sure any misgivings about us will be resolved" [7]. It is as if William Gibson's novel *Idoru* (1996) had come to life. Internet-addiction expert Hiroshi Ashizaki worries, "Today's Japanese youth can't express their true feelings in reality. They can only do it in the virtual world....It's the reverse of reality that they can only talk about what they feel to a friend in the virtual world" [8].[2]

These are like the "techno-virgins" whom Joe Snell envisions in his article "Impacts of Robotic Sex" who do not have "sex with other humans" but see "robotic sex" as "'better' than human sex" [9].

---

[2] Among other sites, pictures are available at Pinoytutorial, "Best and Worst" (Nov 25, 2009), "Man Marries a Video Game – First-ever and the Weirdest," available via http:// pinoytutorial.com/, 2-3.

2.2 Theological Responses

Those who share Dr. Ashizaki's concern include the Internet's Judeo-Christian community. While thorough religious response is not yet plentiful, early reactions include a report from Israel, "Rabbinical Court Debates Status of 'Virtual Sin' in Jewish Law," which notes, "At issue is whether a woman can divorce her husband because he committed virtual adultery using a virtual reality entertainment that is readily available over the Global Landscape." The wife charged, "Insofar as the psychological damage that has been done to me...there is no difference between virtual adultery and the real thing." Her husband, who runs a religious bookstore, responded that he was playing, among others, a game called "King of Israel," that allows one to relive episodes in King David's life, including committing adultery with Bathsheba and murdering Uriah. In his words, "Now, in reliving the life of King David, I am not actually committing adultery, nor am I actually plotting anyone's murder. So, I do not see how my virtual sins, can be compared in nature even to the actual sins of the historic King David, who is considered one of the great Jewish heroes of all time."

Rabbi Aaron Levinsky, Israel's chief rabbi, responded, "King David was a righteous man who repented of his sins. If Mr. Cohen wants to relive the experience of King David he should repent of his sins, abandon these silly entertainments, and devote more time to the study of Torah and religion. I would find Mr. Cohen's argument more convincing if he could compose a virtual Book of Psalms." The rabbi added, "I know that Christians, Moslems, and others are wrestling with these same issues." Princeton Seminary's Sam Humble contributed the Christian view, "Christ established a new standard, to the effect that hatred in the heart was akin to actual murder. Lust in the heart was akin to actual adultery. On that basis, I believe that these new entertainments...promote lust and violence. Thus, I believe that virtual sin is sin." I myself would apply that argument to sex with robots.

A dissenting voice is Rhoda Baker's of the American Civil Liberties Union who cited a Stanford University study that "indicates...committing a virtual murder makes a person less prone to violence," being "a healthy outlet for violent urges" [10].

I, however, think this parallel breaks down, for example, in pederasty, leading to the outlawing of child pornography in many countries due to its ubiquitous presence in cases of sexual molestation of children. There is a process of desensitization (what 1 Timothy 4:2 in the New Testament calls the "searing" of the conscience) and, like all addictions, an escalation that seeks higher and higher stimulation and appears to cross over to real-life referents, in this case children.

Daniel Williams of College Avenue Church of Christ in Arkansas, who holds a Ph.D. in marriage and family therapy, reports, "In my counseling practice, I have already encountered at least a dozen divorces that began with one partner's illicit Internet interactions." He explains, "Sometimes it is the husband who is engaged in these clandestine conversations, but just as often it is the wife" [11].

He, of course, is talking about a computer simulation—but imagine the heightened effect of having never-aging, always compliant, robotic sexual partners in the home. Such a prospect makes me think of the insightful words Isaac Asimov put in the mouth of a developer of a fembot in his short story, "Feminine Intuition": "If women start getting the notion that robots may look like women, I can tell you exactly the kind of perverse notions they'll get, and you'll really have hostility on their part....No woman wants to feel replaceable by something with none of her faults [12].[3]

3 Blurred Lines

As in so many of his predictions, Isaac Asimov was right in this one. As one woman stated what became the consensus opinion of spouses and partners surveyed in studies reported in the journal *Sexual Addiction & Compulsivity*, "[My husband] does not have an actual human mistress from the Internet, but the Internet pornography is the "mistress" that is coming between us. The idealized images of perfect women make me feel inadequate" [13]. Perhaps the 'droids of *Star Wars* are more healthy and appropriate for marital wellbeing than Cherry 2000.

In documenting such responses, a study by the Arizona Community Physicians, from which the previous quotation was drawn, concluded that the reaction of spouses to cybersex extramarital relationships was "hurt, betrayal, rejection, abandonment, devastation, loneliness, shame, isolation, humiliation, jealousy and anger, as well as loss of self esteem" [13]. In fact, every survey reported in the journal's special issue on cybersex revealed feelings of betrayal in spouses or partners.

One man interviewed in the poignant BBC Wonderland Video "Virtual Adultery and Cyberspace Love" had been in a stable relationship for twenty-four years. When his partner asked him point blank, "Are you having an affair with somebody?" he said, "It's a computer game. Don't be silly. It's computer graphics. What could possibly be going on?" He admits, "I felt awful saying that." Soon "it got to the point where I had to do something." He hopes it was the "right thing," but he left his "life's partner" and married in real life the woman with whom he had developed an avatarial relationship. As she explained, "Because we started on Second Life it felt like this is home. People fall in love and get married, so we did" [14].

In a similar manner, a well-programmed malebot or fembot in the house might well alienate a spouse's affection, making one feel, as the Arizona Physicians' survey reported, "unattractive, and even ugly," "sexually rejected, inadequate, and unable to compete" [13].

Such seductive power of a secondary world is subtle, as would be a relationship with a sexbot. As one woman who traded husband and children for a Second Life affair explained, "It kind of hit me before I realized exactly what was happening." "I just kind of fell into it just as though it was natural. And then, every once in a while, I would say—you know, when I would peak a real emotion...what is this? And what am I going to do with it? Ah, and then—ah, I'm not going to think about it. I'm not going to think about that, because this feels good" [14]. Such seductive power caused Emerald Wynn, another resident of Second Life, to avoid all "SL-based romance,"

---

[3] National Geographic reports Hiroshi Ishiguro, who developed the female android Repliee Q1, observing, "When a robot looks too much like the real thing, it's creepy," Chamberlain, Ted (June 23, 2009), Photo in the News: Ultra-Lifelike Robot Debuts in Japan, http://news. National geographic.com/news/2005/06/0610_050610_robot.html, 1.

since "to me, love in Second Life is like a sweet, sweet drug. When I'm with a man I adore there, I feel warm and wrapped in affection....Countless scientific studies show that the mind cannot tell the difference between detailed visualization and reality....Think it and think it hard enough and the mind starts to believe it" [1, 9.2009].

In a similar way, our imagination, with the aid of cleverly designed software, could impute humanity to robots at the expense of the true human with whom we are one flesh. Here I think of the wise words of Le Trung, creator of the femdroid Aiko, "But one thing I will never be able to give her is true emotion or a soul" [15].

4   Implications for Human-Robot Relationships

In light of all this, to me, it seems clear that, in regard to virtual relationships in Second Life, an ethic of consequence should adopt Rabbi Levinsky's repulsion to dwelling virtually on misdeeds and apply Jesus' words about imaginative lust to virtual lust. The view that virtual sex can be pornographic and lethal to one's spiritual health and to one's relationships in the real world, including one's marriage, seems to me indisputable, given the results. So we should conduct ourselves as ethically in the fantasy realm as we do in the real realm.

And, I believe, such information is useful when formulating ethical guidelines for human/robot relationships. Granted, robotics is still in an incipient stage, but, as robotechnology is developing, increasing capabilities are making robots more and more an integral part of human society. This is a good thing. But, as they develop, David Levy, in his thoroughly researched and perceptive book *Love and Sex with Robots: The Evolution of Human-Robot Relationships,* suggests robots will develop to the point that they will become sexual partners of humans. As you can see, that, I believe, would not necessarily be a good thing. In a recently published article, "Should the imago Dei Be Extended to Robots? Love and Sex with Robots, the Future of Marriage, and the Christian Concept of Personhood," available in the free online *Africanus Journal,* [16][4] I consider this question theologically and ethically, exploring various claims about the benefits of developing sexbots as relational partners or even spouses.

Now, should the predictions prove correct and actual robotic sexual partners be introduced into real homes, perhaps in some cases alongside the additional continued presence of virtual partners, what might that mean for our understanding of marriage, faithfulness, adultery, as these issues affect our human relationships, and what kind of adaptive responses will people be suggesting we all make? Here are some possibilities.

4.1   Dispense with Marriage Entirely

Abandon the home as the central, structural unit of society, relegate sex to bonding, entertainment, or exercise, genetically engineer children, and rear them in same-sex

---

[4] Along with being free on the Internet, hard copies of this issue are available at a nominal fee by emailing cumebookstore@gordonconwell.edu, books@gordonconwell.edu, or telephoning CUME bookstore at 617-427-7293 #6207.

dormitories. Such utopian—or dystopian—communities have been envisioned for years. I do not see conservative Judeo-Christian believers embracing this option as a solution.

4.2 Redefine Marriage by Emphasizing Polygamy

This is a biblical reality and, at present, an option in, for example, some African countries where the AIDS crisis has led governments to encourage stable Christian males to take on widowed wives as additional spouses to provide for them, then proceed to educate women, particularly, to adopt a mindset where such behavior is considered normal.

A simple review of the Sarai/Hagar, Rachel/Leah, Hannah/Peninnah conflicts in the Hebrew Bible would, of course, dispel the myth of the happy harem. In my experience as editor of an egalitarian journal, I hear horrendous accounts of polygamy's role in the oppression of women and the validation of patriarchal society. That is why some African churches have chosen to reinstitute the early church's order of widows (see Acts 6:1, 1 Timothy 5:1–17), collectively providing for indigent women who then devote themselves to the ministry of the churches rather than returning to the abuses of polygamy.

4.3 Promote an "Apocalyptic AI" Future

Such a future would involve "outspoken members of the academic community studying the interactions between religion and science," as Robert Geraci states, with what these consider to be a "properly formulated theology," wherein "the image of God means that we form loving relationships with others," suggesting that "the goal of robotics should be the creation of new partners in creation," since "several computer scientists have proposed that the intelligent robots of the future will have religious sentiments" and "some scientists argue that robots will even join humanity in our traditional religious practices and beliefs" [17]. David Levy explains what this would look like: "Whatever the social norms of the prospective owners and their culture, a robot will be able to satisfy them. Similarly with religion, the details and intensity of which can be chosen and changed at will—whether you're looking for an atheist, an occasional church-goer, or a devout member of any religion, you have only to specify your wishes when placing your order at the robot shop" [9]. In this vision, a robot would be like a sophisticated "Chatty Cathy" doll, parroting back to us our faith, as an animated sex and religion doll.

The other way the distinction between human and robot could be diminished is by assuming the Cartesian theory that the human mind can exist separately from the brain and attempting to convert a human's memory banks to electronic form. Should such an assumption be correct and such a process prove possible and able to retain the spirit, or personality, or "pattern of information" [17], and what Henri Bergson called the *élan vital,* the creative force within us with which we adapt and grow, humans might download our minds either into a Second Life–type avatar and, after our human bodies die, as in the vision of Vernor Vinge in *True Names* (1981),[5] live forever digitally, or at least as long as the virtual world is maintained, or, in the real world, download our minds into a robot's mechanical brain in order to prolong our lives as cyborgs.[6] In this robotic version, the hardware or cyborg form, we might argue humans essentially remain ourselves, since, along with a mechanical knee, artificial hip, pacemaker, all common today, we would include a brain, but what makes us ourselves—that which disappears so utterly at death—would remain, within the body. I suggest that after consideration many devout Jews or Christians might accept this development, concluding that God created an original world of raw material and gifted each human with a unique mind and spirit and some of us with the capability to improve all our lives by using that material constructively. But, becoming cyborgs would not make us androids. Our origins would still differ.

In that sense, the nature of the moral universe in which we exist does not change with technology, as G.K. Chesterton counseled through the mouth of his fictional "avatar," Father Brown:

> Reason and justice grip the remotest and the loneliest star...you can imagine any mad botany or geology you please. Think of forests of adamant with leaves of brilliants. Think the moon is a blue moon, a single elephantine sapphire. But don't fancy that all that frantic astronomy would make the smallest difference to the reason and justice of conduct. On plains of opal, under cliffs cut out of pearl, you would still find a notice-board, "Thou shalt not steal." [18]

Someone else's spouse, either virtually or in reality, I would add.

4.4 Maintain the Distinctions, but with Respect

I believe a workable ethic would be one that values the differences between humans and robots and treats marriage as an enduring human institution. While we value robots as worthy works of our hands, made in our human image, as we are in turn made in God's image, and gifted by us to exist in cooperative relationships, they should not be used in marital or sexual ones. I would draw the same conclusion for participation with an avatar of a machine or even of other humans in digital worlds with whom one is not married in the real world. Technology has not replaced the Ten Commandments, in my estimation, and I do not see why a more sophisticated form of it would.

In conclusion, in the case of extramarital virtual (or, by inference, robotic) sex, particularly, a cross-section of thinkers have already observed it is pain-producing and

---

[5] The idea of an image that outlives its human referent as a separate entity, although without self-determination, was explored that same year (1981) in the film, "Looker." In 2002's "Simone," a movie producer appears to fall in love with a composite image, which he has constructed digitally and which has become a simulated star performer in its own right.

[6] Robert Geraci (2010, 32) credits Hans Moravec's 1978 *Analog* article, "Today's Computers, Intelligent Machines and Our Future," with applying this idea to religion: "Nearly all of the Apocalyptic AI advocates agree that human beings will eventually learn to 'upload' a human mind into a robot body...a position first advocated by Moravec in *Analog*."

wrong. I do not think that promoting it technologically would enhance the future of our descendants' inter-human or human/robot relationships.[7]

References

1. Au, W.: New World Notes (2006-2010) Snowcrashed (August 14, 2006), `http://nwn.blogs.com`; If virtual adultery is so common, why does the media keep rehashing the same virtual adultery stories? (February 2009); Sex and romance, entries (June 30, August 7, August 20, September 9, September 11, November 17, 2009); Philip Rosedale attempting to create sentient artificial intelligence that thinks and dreams in Second Life! (February 3, 2010)
2. Stephenson, N.: Snow Crash, pp. 33–34. Bantam, New York (1993)
3. Linden, M.: Avatars unite! (January 29, 2010), `https://blogs.secondlife.com/community/features/blog/2010/01/29/avatars-unite`
4. Young, K.: Has your relationship been hurt by a cyberaffair? (December 13, 2008), `http://www.healthyplace.com/addictions/center-for-internet-addiction-recovery/has-your-relationship-been-hurt-by-a-cyberaffair/menu-id-1105`
5. Alter, A.: Is this man cheating on his wife? Alexandra Alter on the toll one man's virtual marriage is taking on his real one and what researchers are discovering about the surprising power of synthetic identity (August 10, 2007), `http://online.wsj.com/article/SB11867016455923/93622.html`
6. Pothen, J.: Virtual reality? Speculation on sex, divorce and cyberspace (December 2, 2008), `http://cornellsun.com/node/33953`
7. Moses, A.: Man marries virtual girlfriend (November 26, 2009), `http://www.stuff.co.nz/technology/digital-living/3100430/Man-marries-virtual-girfriend`
8. Lah, K.: Tokyo man marries video game character (December 17, 2009), `http://www.cnn.com/2009/WORLD/asiapcf/12/16/japan.virtual.wedding/index.html`
9. Levy, D.: Love and Sex with Robots: The Evolution of Human-Robot Relationships, p. 138. HarperCollins, New York (2007)
10. Aharoni, Y.: Quandary in Israel: rabbinical court debates status of "virtual sin" in Jewish law (1999), `http://www.cs.wcupa.edu/epstein/rabbinic.htm`
11. Williams, D.: Virtual adultery, N.D., `http://www.cacoc.org/virtual_adultery,htm`

---

[7] The need to resolve such issues is important because the future promises to bring us even more complex ones to process. Recently, Philip Rosedale gave more details about his diminished involvement in Second Life in favor of his work in LoveMachine Inc., a new company he organized with Ryan Downe. According to Hikaru Yamamoto, he is "creating a sentient artificial intelligence" to exist "in a virtual world. 'He wants it to live inside Second Life...It will think and dream and everything,'" predicating on the question: "Can 10,000 computers become a person?" Philip Rosedale himself admitted to Wagner James Au, "[T]hat's the general direction we're going" (Au 3.2010, 9). Obviously, humans are facing large ethical issues in the not-too-distant future, in this case in the area of personhood. We need to work together to set cooperative policies as new issues face us.

12. Asimov, I.: "Feminine Intuition" in Bicentennial Man and Other Stories. Doubleday, Garden City, N.Y., p. 9, 12 (1976)
13. Schneider, J.: Effects of Cybersex Addiction on the Family: Results of a Survey. In: Cooper, A.l. (ed.) Cybersex: The Dark Side of the Force, A Special Issue of the Journal Sexual Addiction and Compulsivity, p.31,38,41,46s. Routledge, New York (2000)
14. Wonderland: Virtual adultery and cyberspace love, video broadcast, BBC Two (UK) (January 30, 2008), http://todocumentaryfilms.com/virtual-adultery/
15. Trung, L.: About Aiko: Past, Present, and Future: NO perfection..., http://www.projectaiko.com/about.html
16. Spencer, W.D.: Should the imago Dei Be Extended to Robots? Love and Sex with Robots, the Future of Marriage, and the Christian Concept of Personhood. Africanus Journal 1(2), 6–19, http://www.gordonconwell.edu/boston/africanus_journal
17. Geraci, R.M.: Apcalyptic AI: Visions of Heaven in Robotics, Artificial Intelligence and Virtual Reality, p. 5. Oxford University Press, New York (2010)
18. Chesterton, G.K.: The Blue Cross, in The Innocence of Father Brown (1911); Collected in The Father Brown Omnibus, Dodd, Mead, New York, p. 19 (1951)

Modeling Mixed Groups of Humans and Robots with Reflexive Game Theory

Sergey Tarasenko

Department of Intelligence Science and Technology, Graduate School of Informatics
Kyoto University, Yoshida honmachi, Kyoto 606-8501, Japan
infra.core@gmail.com

Abstract. The Reflexive Game Theory is based on decision-making principles similar to the ones used by humans. This theory considers groups of subjects and allows to predict which action from the set each subject in the group will choose. It is possible to influence subject's decision in a way that he will make a particular choice. The purpose of this study is to illustrate how robots can refrain humans from risky actions. To determine the risky actions, the Asimov's Three Laws of robotics are employed. By fusing the RGT's power to convince humans on the mental level with Asimov's Laws' safety, we illustrate how robots in the mixed groups of humans and robots can influence on human subjects in order to refrain humans from risky actions. We suggest that this fusion has a potential to device human-like motor behaving and looking robots with the human-like decision-making algorithms.

Keywords: Reflexive Game Theory (RGT), Asimov's Laws of Robotics, mixed groups of humans and robots, human-robot societies.

1 Introduction

Now days robots have become an essential part of our life. One of the purposes robots serve to is to substitute human beings in dangerous situations, like defuse a bomb etc. However human nature shows strong inclinations towards the risky behavior, which can cause not only injuries, but even threaten the human life. The list of these reasons includes a wide range starting from irresponsible kids' behavior to necessity to find solution in a critical situation. In such a situation, a robot should full-fill a function of refraining humans from doing risky actions and perform the risky action itself.

However, robot should not physically force people, but must convince people on the mental level to refrain from doing an action. This method is more effective rather than a simple physical compulsion, because humans make the decisions themselves and treat these decisions as their own. This approach is called a *reflexive control* [1].

We consider the mixed groups of humans and robots. To be able to interact with humans on the mental level robot should posses an ability to "think" and make decisions in a way similar to the one that humans have.

M.H. Lamers and F.J. Verbeek (Eds.): HRPR 2010, LNICST 59, pp. 108–117, 2011.

The principles explaining human decision-making are the basis of the Reflexive Game Theory (RGT) proposed and developed by Lefebvre [1,2,3]. By using the RGT, it is possible to predict choices made by each individual in the group and influence on their decision-making. In particular, the RGT can be used to predict terrorists' behavior [4].

The purpose of the present study is to apply RGT for analysis of individual's behavior in the mixed groups of humans and robots and illustrate how RGT can be used by robots to refrain humans from doing risky actions. We start with brief description of the RGT and illustrate its application with a simple example. Then we formalize the definition of robots, distinguishing them from humans. Finally, we consider two examples, in which humans tend to do risky actions, and show how robots, using RGT, can refrain humans from doing these actions.

2   Brief Overview of the Reflexive Game Theory (RGT)

The RGT deals with groups of individuals (subjects, agents etc). Any group of subjects is represented in the form of *fully connected graph*. In the present study, a subject can be either a human or a robot. Each subject is assigned a unique variable (*subject variable*), which is a vertex of the graph. The RGT uses the set theory and Boolean algebra as the basis for calculus. Therefore the values of subject variables are elements of Boolean algebra.

All the subjects in the group can have either alliance or conflict relationship. The relationships are identified as a result of group macroanalysis. It is suggested that the installed relationships can be changed. The relationships are illustrated with graph ribs. The solid-line ribs correspond to alliance, while dashed ones are considered as conflict. For mathematical analysis alliance is considered to be conjunction (multiplication) operation ($\cdot$), and conflict is defined as disjunction (summation) operation ($+$).

The graph presented in Fig. 1a or any graph containing any sub-graph isomorphic to this graph are not decomposable. In this case, the subjects are excluded from the group one by one, until the graph becomes decomposable. The exclusion is done according to importance of the other subjects for a particular one. Any other fully connected graphs are decomposable and can be presented in an analytic form of a corresponding *polynomial*. Any relationship graph of three subjects is decomposable (see [3,4]).

Consider three subjects a, b and c. Let subject a is in alliance with other subjects, while subjects b and c are in conflict (Fig. 1b). The polynomial corresponding to this graph is $a(b + c)$.

Regarding the relationship, the polynomial can be stratified (decomposed) into *sub-polynomials* [2,3,4,5]. Each sub-polynomial belongs to a particular level of stratification. If the stratification regarding alliance was first built, then the stratification regarding the conflict is implemented on the next step. The stratification procedure finalizes, when the *elementary polynomials*, containing a single variable, are obtained after a certain stratification.

The result of stratification is the *polynomial stratification tree (PST)*. It has been proved that each non-elementary polynomial can be stratified in an unique

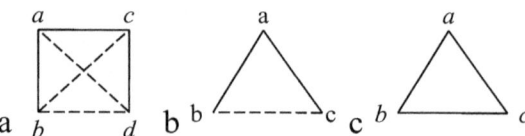

Fig. 1. The relationship graphs

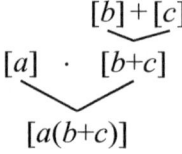

Fig. 2. Polynomial Stratification Tree. Polynomials $[a], [b]$ and $[c]$ are elementary polynomials.

way, i.e., each non-elementary polynomial has only one corresponding PST [5]. Each higher level of the tree contains polynomials simpler than the ones on the lower level. For the purpose of stratification the polynomials are written in square brackets. The PST for polynomial $a(b + c)$ is presented in Fig. 2.

Next, we omit the branches of the PST and from each non-elementary polynomial write in top right corner its sub-polynomials. The resulting tree-like structure is called a *diagonal form*. Consider the diagonal form corresponding to the PST in Fig. 2:

$$[a(b+c)] \quad [a][b+c] \quad \overset{[b]+[c]}{} \quad .$$

Hereafter the diagonal form is considered as a function defined on the set of all subsets of the *universal set*. The universal set contains the *elementary actions*. For example, these actions are actions α and β. The *Boolean algebra* of the universal set includes four elements: $1 = \{\alpha, \beta\}, \{\alpha\}, \{\beta\}$ and the empty set $0 = \{\}$. These elements are all the possible subsets of universal set and considered as alternatives that each subject can choose. The alternative $0 = \{\}$ is interpreted as an inactive or idle state. In general, Boolean algebra consists of 2^n alternatives, if universal set contains n actions.

Formula $P^W = P + \overline{W}$, where $\overline{W}$ stands for negation of W [6], is used further to fold the diagonal form. During the folding, round and square brackets are considered to be interchangeable. The following equalities are also considered to be true: $x + \overline{x} = 1, x + 0 = x$ and $x + 1 = 1$. Next we implement folding of diagonal form of polynomial $a(b + c)$:

$$[a(b+c)] \quad [a][b+c] \quad \overset{[b]+[c]}{} = [a(b+c)] \quad [a]([b+c] + \overline{[b]+[c]}) = a(b+c) + \overline{a} .$$

The goal of each subject in the group is to choose an alternative from the set of alternatives under consideration. To obtain choice of each subject, we consider the *decision equations*, which contain subject variable in the left-hand side and the result of diagonal form folding in the right-hand side: $a = (b + c)a + \bar{a}, b = (b + c)a + \bar{a}$ and $c = (b + c)a + \bar{a}$.

To find solution of the decision equations, we consider the following equation, which is a *canonical form of the decision equation*:

$$x = Ax + B\bar{x} , \qquad (1)$$

where x is the subject variable, and A and B are some sets. This equation has solution if and only if the set B is contained in set A: $A \supseteq B$. If this requirement is satisfied, then equation has at least one solution from the interval $A \supseteq x \supseteq B$ [6]. Otherwise, the decision equation (1) has no solution, and it is considered that subject cannot make a decision. Thus, he is considered to be in frustration state.

Therefore, to find solutions of equation, one should first transform it into the *canonical form*. Out of three presented equations only the decision equation for subject a is in the canonical form, while other two should be transformed into. We consider explicit transformation only of decision equation for subject b: $a(b+c)+\bar{a} = ab+ac+\bar{a} = ab+(ac+\bar{a})b+(ac+\bar{a})\bar{b} = (a+\bar{a}+ac)b+(ac+\bar{a})\bar{b} = (1+ac)b + (ac + \bar{a})\bar{b} = b + (ac + \bar{a})\bar{b}$. Therefore, $b = b + (ac + \bar{a})\bar{b}$.

The transformation of equation for subject c be can be easily derived by analogy: $c = c + (ab + \bar{a})\bar{c}$.

Table 1. Influence Matrix

	a	b	c
a	a	$\{\alpha\}$	$\{\beta\}$
b	$\{\beta\}$	b	$\{\beta\}$
c	$\{\beta\}$	$\{\beta\}$	c

The variable in the left-hand side of decision equation in the canonical form is the variable of the equations, while other variables are considered as influences on the subject from the other subjects. All the influences are presented in the *Influence matrix* (Table 1). The main diagonal of the Influence matrix contains the subject variables. The rows of the matrix represent influences of the given subject on other subjects, while columns represent the influences of other subjects on the given one. The influence values are used in decision equations.

For subject a: $a = (\{\beta\} + \{\beta\})a + \bar{a} \Rightarrow a = \{\beta\}a + \bar{a}$.
For subject b: $b = b + (\{\alpha\}\{\beta\} + \overline{\{\alpha\}})\bar{b} \Rightarrow b = b + \{\beta\}\bar{b}$.
For subject c: $c = c + (\{\beta\}\{\beta\} + \overline{\{\beta\}})\bar{c} \Rightarrow c = c + (\{\beta\} + \{\alpha\})\bar{c} \Rightarrow c = 1$.

Equation for subject a does not have any solutions, since set $A = \{\beta\}$ is contained in set $B = 1$: $A \subset B$. Therefore, subject a cannot make any decision and is in frustration state.

Equation for subject b has at least one solution, since $A = 1 = \{\alpha, \beta\} \supseteq B = \{\beta\}$. The solution belongs to the interval $1 \supseteq b \supseteq \{\beta\}$. Therefore subject b can choose any alternative from Boolean algebra, which contains alternative $\{\beta\}$. Thus, only alternative $\{\beta\}$ can be implemented.

Equation for subject c turns into equality $c = 1$. This is possible only in the case, when $A = B$. Here $A = B = 1$. Subject c can implement any alternative except for alternative $0 = \{\}$. However, he does not have absolute freedom of choice, since this implies ability to choose inactive alternative $0 = \{\}$, as well.

This concludes a brief overview of the RGT. Next we consider formalization of robotic subjects.

3   Defining Robots in RGT

It is considered by default that robot follows the program of behavior generated by the control system. This control system consists of at least three modules. The Module 1 implements robot's ability of human-like decision-making. The Module 2 contains the rules, which refrain robot from making a harm to human beings. The Module 3 predicts the choice of each human subject and suggests the possible strategies of reflexive control.

We suggest to apply Asimov's Three Laws of robotics [7], which formulate the basics of the Module 2:

1) a robot may not injure a human being or, through inaction, allow a human being to come to harm;
2) a robot must obey any orders given to it by human beings, except where such orders would conflict with the First Law;
3) a robot must protect its own existence as long as such protection does not conflict with the First or Second Law.

We consider that these Laws are intrinsic part of robots "mind", which cannot be erased or corrupted by any means.

The interaction of Modules 1, 2 and 3 in robot's control system is presented in Fig. 3. First the information from environment is formalized as the Boolean algebra of possible alternatives. Then the human-like decision-making system is implemented in Modules 1 and 3. The robot's decision-making based on the RGT is implemented in Module 1. The output of Module 1 is set D, which contains solution of robot's decision equation.

The Boolean algebra is filtered according to Asimov's Laws in Module 2. The output of Module 2 is set U of approved alternatives. Then the conjunction of sets D and U is performed: $D \cap U = DU$. If DU is the empty set ($DU = \{\}$), then a robot chooses alternative from the set U. If set DU is not empty, then a robot selects the actions from the set DU.

To achieve the goal of refraining human subjects from risky action, robot predicts choice of each human subject in Module 3. The decision-making system similar to the one in Module 1 is employed. The output set D_h corresponds to output set D. If robot predicts that choice of some human subject is risky alternative, the robot analyzes all the possible scenarios which succeed in not

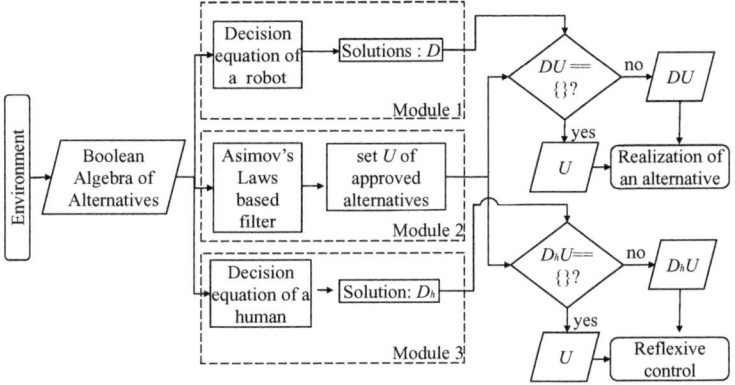

Fig. 3. Schematic representation of robot's control system

choosing the risky alternative and applies reflexive control according to this scenario.

The result of interaction between Modules 1 and 2 includes robot's choices, which are harmless for human subjects, while result of interaction between Modules 3 and 2 is the set of possible scenarios of reflexive control. This is important difference between the outputs of Modules 1 and 2 and Modules 3 and 2.

4 Sample Analysis of Mixed Groups

Here we consider two examples of how robots in the mixed groups can make humans refrain from risky actions. The first example considers robot baby-sitters. The second one illustrates the critical situation with mountain-climbers and rescue robot. The goal of robots in both situations is to refrain humans from doing risky actions.

4.1 Robot Baby-Sitters

Suppose robots have to play a part of baby-sitters by looking after the kids. We consider a mixed group of two kids and two robots. Each robot is looking after a particular kid. Having finished the game, kids are considering what to do next. They choose between to compete climbing the high tree (action α) and to play with a ball (action β). Together actions α and β represent the active state $1 = \{\alpha, \beta\} = \{\alpha\} + \{\beta\}$. Therefore the Boolean algebra of alternatives consists of four elements: 1) the alternative $\{\alpha\}$ is to climb the tree; 2) the alternative $\{\beta\}$ is to play with a ball; 3) the alternative $1 = \{\alpha, \beta\}$ means that a kid is hesitating what to do; and 4) the alternative $0 = \{\}$ means to take a rest.

We consider that each kid considers his robot as ally and another kid and his robot as the competitors. The kids are subjects a and c, while robots are subjects b and d. The relationship graph is presented in Fig. 4.

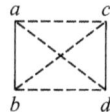

Fig. 4. The relationship graph for robot baby-sitters example

Next we calculate the diagonal form and fold it in order to obtain decision equation for each subject:

$$
\begin{array}{c}
[a][b] \qquad\qquad [c][d] \\
[ab] \qquad\qquad +[cd] \\
[ab + cd] \qquad\qquad\qquad\qquad\qquad = ab + cd\ .
\end{array}
$$

From two actions α and β, action α is a risky action, since a kid can fall from the tree and this is real threat for his health or even life. Therefore according to Asimov's Laws, robots cannot allow kids to start the competition. Thus, robots have to convince kids not to choose alternative $\{\alpha\}$. In terms of alternatives, the Asimov's Laws serve like filters, which excludes the risky alternatives. The remaining alternatives are included into set U. In this case, $U = \{\{\beta\}, \{\}\}$.

Next we discuss some possible scenarios.

Scenario 1. Let every robot tries to convince the kid, it is looking after, to play with a ball, i.e., $b = d = \{\beta\}$: $a = a\{\beta\} + c\{\beta\}$ and $c = a\{\beta\} + c\{\beta\}$.

Then there are the following solutions for kid a.

If $c\{\beta\} = 0$, then $a = a\{\beta\}$. This equation has two solutions 0 and $\{\beta\}$. The equality $c\{\beta\} = 0$ holds, if c equals 0 or $\{\alpha\}$. Therefore, if both robots influence $\{\beta\}$ and kid c influences either 0 or $\{\alpha\}$ on kid a, kid a can choose either to have a rest or to play a ball.

If $c\{\beta\} \neq 0$, then $c\{\beta\} = \{\beta\}$ and $a = a\{\beta\} + \{\beta\} = (a + 1)\{\beta\} = \{\beta\}$. The equality $c\{\beta\} = \{\beta\}$ holds, if c equals either 1 or $\{\beta\}$. Therefore, if both robots influence $\{\beta\}$ and kid c influences either 1 or $\{\beta\}$ on kid a, kid a can only choose to play a ball.

The kid c will behave in the same way, if we exchange the roles of kid a and kid c.

Scenario 2. Both robots influence $0 = \{\}$ on both kids. In this case, to have the rest is the only option for the kids: $ab + cd = a0 + c0 = 0 \Rightarrow a = 0$ and $c = 0$.

In the presented example, robots can successfully control kids' behavior by refraining them from doing risky actions.

4.2   Mountain-Climbers and Rescue Robot

We consider that there are two climbers and rescue robot in the mountains. The climbers and robot are communicating via radio. One of the climbers (subject b) got into difficult situation and needs help. Suggest, he fell into the rift because the edge of the rift was covered with ice. The rift is not too deep and there is a thick layer of snow on the bottom, therefore climber is not hurt, but he cannot get out of the rift himself. The second climber (subject a) wants to rescue his friend himself (action α), which is risky action. The second option is that robot will perform rescue mission (action β). In this case, the set U of approved alternatives for robot includes only alternative $\{\beta\}$, since inaction is inappropriate solution according to the First Law. The goal of the robot is to refrain the climber a from choosing alternative $\{\alpha\}$ and perform rescue mission itself. We suggest that from the beginning all subjects are in alliance. The corresponding graph is presented in Fig. 1c and its polynomial is abc.

Next we calculate diagonal form and perform folding procedure:

$$\frac{[a][b][c]}{[abc]} = [abc] + \overline{[a][b][c]} = 1 .$$

Thus, any subject in the group is in active state. Therefore, group is uncontrollable. In this case, robot makes decision to change his relationship with the climber b from alliance to conflict. Robot can do that, for instance, by not responding to climber's orders. Then the relationship graph transforms into the graph depicted on Fig. 1b and the decisions of subjects are defined by the following equations: $a = (b + c)a + \overline{a}; b = b + (ac + \overline{a})\overline{b}$ and $c = c + (ab + \overline{a})\overline{c}$ (see section 2).

The choice of climber a is defined by the interval $(b + c) \supseteq a \supseteq 1$. Therefore, climber a is capable of choosing only alternative $1 = \{\alpha, \beta\}$, if the condition $(b+c) = 1$ is satisfied. Thus, next climber a can realize one of the alternatives $\{\alpha\}$ or $\{\beta\}$. This case is not acceptable, since climber a can realize risky alternative $\{\alpha\}$. On the other hand, if $(b+c) \subset 1$, then climber a is in frustration and cannot make any choice.

Therefore the only way to refrain climber a from choosing alternative $\{\alpha\}$ is to put him into frustration state.

Next we consider various options of climber's b influence on climber a. Let at first he makes influence to send the robot (alternative $\{\beta\}$). In this case, if $(b + c) \subset 1$, then climber a gets into frustration and cannot make any decision. Therefore the influence of robot c on the climber a should be $\{\beta\}$, as well. Then $(b + c) = (\{\beta\} + \{\beta\}) = \{\beta\} \subset 1$, and climber a cannot make any decision.

If climber b makes influence $\{\alpha\}$ on climber a, then robot has to make influence $\{\alpha\}$ on climber a, as well. Then $(b + c) = (\{\alpha\} + \{\alpha\}) = \{\alpha\} \subset 1$, and climber a cannot make a decision.

Next, we illustrate that regardless of climbers' simultaneous (joint) influences on the robot, it can realize alternative $\{\beta\}$, thus, completing the rescue mission.

Here four scenarios of climbers' joint influences on the robot c are considered.

Scenario 1. Climbers a and b make influences $\{\alpha\}$ and $\{\beta\}$, respectively. Then $a = \{\alpha\}, b = \{\beta\}$: $1 \supseteq c \supseteq \{\alpha\}\{\beta\} + \overline{\{\alpha\}} \Rightarrow 1 \supseteq c \supseteq \{\beta\}$. Therefore robot can choose any alternative, which includes alternative $\{\beta\}$. In this case, $D = \{\{\alpha, \beta\}, \{\beta\}\}$ and $U = \{\{\beta\}\}$, consequently, $DU = \{\{\beta\}\}$. Therefore robot will choose alternative $\{\beta\}$.

Scenario 2. Climbers a and b make influences $\{\beta\}$ and $\{\alpha\}$, respectively. Then $a = \{\beta\}, b = \{\alpha\}$: $1 \supseteq c \supseteq \{\beta\}\{\alpha\} + \overline{\{\beta\}} \Rightarrow 1 \supseteq c \supseteq \{\alpha\}$. Therefore robot can choose any alternative, which includes alternative $\{\alpha\}$: $D = \{\{\alpha, \beta\}, \{\alpha\}\}$. Since $U = \{\{\beta\}\}$, $DU = \{\}$. According to the control schema in Fig. 3, robot will be choosing from alternatives in set U. Therefore, robot will choose alternative $\{\beta\}$.

Scenario 3. Both climbers make influences $\{\alpha\}$. Then for robot c, $a = b = \{\alpha\}$: $1 \supseteq c \supseteq \{\alpha\}\{\alpha\} + \overline{\{\alpha\}} \Rightarrow c = 1$ and $D = \{\{\alpha, \beta\}\}$. Since $U = \{\{\beta\}\}$, $DU = \{\}$. Thus as in the previous scenario, robot will choose alternative $\{\beta\}$.

Scenario 4. Both climbers make influences $\{\beta\}$. Then for robot c, $a = b = \{\beta\}$: $1 \supseteq c \supseteq \{\beta\}\{\beta\} + \overline{\{\beta\}} \Rightarrow c = 1$ and $D = \{\{\alpha, \beta\}\}$. Since $U = \{\{\beta\}\}$, $DU = \{\}$ and robot will choose alternative $\{\beta\}$.

The discussed example illustrates how robot can transform uncontrollable group into controllable one by manipulating the relationships in the group. In the controllable group by its influence on the human subjects, robot can refrain the climber a from risky action to rescue climber b. Robot achieves its goal by putting climber a into frustration state, in which climber a cannot make any decision. On the other hand, set U of approved alternatives guarantees that robot itself will choose the option with no risk for humans and implement it regardless of climber's influence.

5   Discussion and Conclusion

In the present study we have shown how the Reflexive Game Theory merged with Asimov's Laws of robotics can enable robots to refrain the human beings from doing risky actions. The beauty of this approach is that subjects make decisions themselves and consider these decisions as their own. Therefore the RGT fused with Asimov's Laws play a part of social buffer, providing safer life for people with no extensive psychological pressure on human subjects. To our knowledge up to date there has been no similar approach proposed.

The first example illustrates how robots can filter out the risky action and choose both active and inactive alternatives. Here robots are not required to perform any actions. The second example describes critical conditions. In this case, the inactive alternative cannot be chosen and robot has to take the burden of performing risky action. The example with kids illustrates less intense conditions of robot's inference application, while the second example requires more sophisticated multistage strategies: 1) to change the group's configuration; 2) to refrain a climber from trying to go on a rescue mission himself; and 3) to perform the rescue mission by robot itself. The first example plays a role of introductory passage from the kids' yard games to the real life situation in the mountains, which requires more powerful calculus than the first one. The proposed approach based on fusion of the RGT with Asimov's Laws shows its capability of managing successfully the robots' behavior in either situation.

This allows to make the next step in human and robot integration. The RGT provides human-like decision-making system, thus enabling robots to track the human decisions and influence on the people in the way humans are perceptive to. The RGT presents the formal and general model of the group. This model enables to compute the choices of human subjects in the form of feasible algorithms and apply the reflexive control accordingly. Therefore, the RGT is the core part enabling robots to "think" like humans. The Asimov's Laws are the basis for filtering out the risky actions from the existing alternative. Thus, robots programmed with fusion of the RGT and Asimov's Laws are the tools to create risk free psychologically friendly environments for human beings. It opens prospectives of creating the robotic agent, capable of behaving itself like human beings on both motor and mental levels. The primary goal the robots thought to be used for was to substitute the human beings in deadly environments or in the case, when human abilities are not enough, for instance, when extremely high precision is needed. The capability of "thinking" like humans opens new frontiers to the outer limits of the human nature.

The core idea of this study is to show how human-like robots can refrain human beings from doing the risky actions by using the RGT and Asimov's Laws. Therefore the questions of development of required interfaces to extract Boolean algebra directly from the environment are not discussed. The present study also is not answering the technical issues as software and hardware implementation. We consider these questions as the future research trends. This study only shows how human mental world can be represented in the form of feasible RGT algorithms, then fused with Asimov's Laws and implanted into robots' mind. We hope the result of this fusion is a one small step towards making our life safer and our world a better place.

References

1. Lefebvre, V.A.: The basic ideas of reflexive game's logic. Problems of systems and structures research, 73–79 (1965) (in Russian)
2. Lefebvre, V.A.: Lectures on Reflexive Game Theory. Cogito-Centre, Moscow, (2009) (in Russian)
3. Lefebvre, V.A.: Lectures on Reflexive Game Theory. Leaf & Oaks, Los Angeles (2010)
4. Lefebvre, V.A.: Reflexive analysis of groups. In: Argamon, S., Howard, N. (eds.) Computational models for counterterrorism, pp. 173–210. Springer, Heidelberg (2009)
5. Batchelder, W.H., Lefebvre, V.A.: A mathematical analysis of a natural class of partitions of a graph. J. Math. Psy. 26, 124–148 (1982)
6. Lefebvre, V.A.: Algebra of Conscience. D. Reidel, Holland (1982)
7. Asimov, I.: Runaround. Astounding Science Fiction, 94–103 (March 1942)

A Design Process for Lovotics

Hooman Aghaebrahimi Samani[1,2], Adrian David Cheok[1,2],
Mili John Tharakan[2], Jeffrey Koh[1,2], and Newton Fernando[2]

[1] Graduate School for Integrative Science and Engineering,
National University of Singapore
[2] Keio-NUS Cute Center,
National University of Singapore and Keio University, Japan
{hooman,elecad,idmmjt,g0901851,idmfonn}@nus.edu.sg

Abstract. We refer to human-robot relationships as Lovotics. In this
paper a design process for Lovotics is presented. In order to invoke these
relationships, technological solutions can only take us so far. Design
played an important role in order to engage users to explore the possibilities of bi-directional, human-robot love. We conducted a user-centric
study in order to understand these factors and incorporate them into
our design. The key issues of design for developing a strong emotional
connection between robots and humans are investigated. A questionnaire
is proposed and based on the results of this a robot with minimal design
is developed.

1 Introduction

Lovotics is a research domain for developing a love-like relationship between
humans and robot. This multidisciplinary research investigates various scientific issues regarding human-robot love. One of the primary requirements of the
Lovotics research is to design a robot that loves and is loved by humans. "Love"
is an abstract term. The first and most trying difficulty regarding our project
was to outline a definition for the very emotion we wanted to simulate. Because
of this, we decided to begin with a textbook definition of the term in order to
frame our research. If we could recreate this emotion within a robot at its basic definition, we felt that we could then begin to design new definitions of the
term which included robots in the picture. As such, we have used the following
definition by Aristotle as a starting point:

> "Philia (φιλία), a dispassionate virtuous love, was a concept developed
> by Aristotle [1]. It includes loyalty to friends, family, and community, and
> requires virtue, equality, and familiarity. Philia is motivated by practical
> reasons; one or both of the parties benefit from the relationship. It can
> also mean love of the mind [2]. Philia entails a fondness and appreciation
> of the other. For the Greeks, the term Philia incorporated not just friend-
> ship, but also loyalties to family and polis-ones political community, job,
> or discipline [3]."

M.H. Lamers and F.J. Verbeek (Eds.): HRPR 2010, LNICST 59, pp. 118–125, 2011.

2   System Design

Lovotics is the beginning of a series of work that explores the notion of bi-directional Human - Robot love. We see this mutual relationship and love as the future of robotics, where we envision a world as a place where humans have robots as friends and possibly even life partners. To achieve this relationship, we developed Lovotics, a love robot that aims to capture the essence of a Philia relationship and translate that for robotics to enable this relationship to grow between robots and humans.

The results of our questionnaire show varied responses to how someone expresses or feels love. We have taken a multidisciplinary approach to understand the basic elements to 'love' and recreated this in Lovotics robot. We have identified the senses - tactile, audio and visual, as the basic key senses used for expressing and feeling love.

Donald Norman introduced three levels of processing in design - visceral, behavioral and reflective [4]. These three levels of design are inter-related and the success of a product is usually determined by how good the design is at these three levels.

I. Visceral design is dominated by how products look, feel and even sound. The underlying principles are usually hard-wired in, consistent across people and culture. For Lovotics, there is a need for it to encourage sociability with human being and thus its appearance is key as well as its tactile, audio and visual input and output.

II. Reflective design is about what the product means to the user, the message it brings across and its cultural impact. For Lovotics, the meaning will be to create a robot that loves the user and evoke feelings of love from him/her and the vision is to create a culture of love with robots, changing the human perception of robot beings without feeling.

III. Behavior design is about how the product performs. Good behavior design has four main components: function, understandability, usability and physical feel. For Lovotics, its main aim is to infer sociability with humans through love with interactions that are comprehensible and intuitive.

Affordance is a psychological terms that means giving a message from the object [5]. It defines the intuitiveness of the object and makes it easier to understand where the function of the product is. When the affordance of an object or environment corresponds with its intended function, the design will perform more efficiently and will be easier to use. Lovotics robots need to immediately give the message to the user of its function and purpose.

The uncanny valley [6] needs to be avoided and a more simplified representations of characters for the robot may be more acceptable to the human brain than explicitly realistic representations.

3   Pre-evaluation Survey

Before making any attempt to design a robot which could accurately communicate the feeling of love, we needed to get a better understanding of how people perceive love and objects. In order to do so we wanted to first pose questions that would instigate memories of love and being loved from a human-to-human context. Focusing on personal perspective, we designed questions that would engage the participant's previous experiences in order to prepare them for our next set of questions.

In the second half of the survey, we wanted to get responses regarding whether or not people could get comfortable with the idea of loving robots. These questions mostly dealt with whether they could imagine loving a robot, list some of their favorite objects and what words they would use to define love with objects. These questions helped us define a behavioral model when designing the robot.

Lastly, we wanted to get a more concrete understanding of the embodiment of physical qualities within an object when imagining a robot that could love and be loved. These questions helped us define a physical design for the robot, and also helped us define other modes of expression that the robot needed in order to engage with real humans.

The following are results from a survey completed by 70 participants of various age, race, occupation and sex. The survey was conducted over the internet and participants were found through various social networking channels.

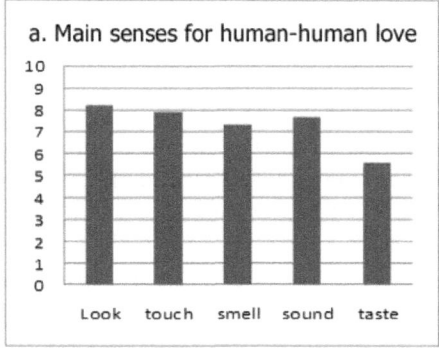 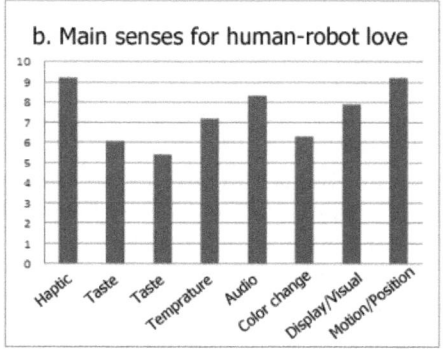

Fig. 1. Questionnaire results for main sensory channels for Lovotics. a. Modalities for human-human love b. Modalities for human-robot love.

From our pool of participants, less then half could not imagine loving a robot. A majority of the participants though were at least open to the idea, with 19% of participants actually believing that they could love a robot. These answers contrasted somewhat when the pool of participants were asked if they could in turn be loved by robots, with 77% open to being loved by robots. People are more willing to be loved and then to love, it seems.

Survey participants valued appearance above all sensations in regards to human-to-human love. Touch and sound came second in importance regarding human-to-human love, with smell a close third and taste being least important. Results from the survey discussing human-to-robot love pointed towards similar

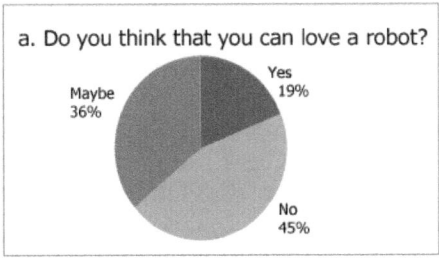
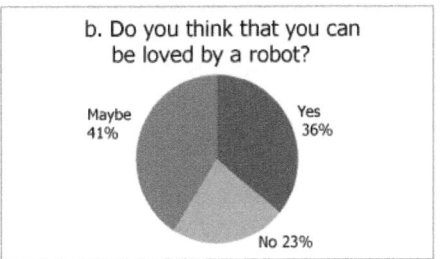

Fig. 2. Questionnaire results for human-robot love imagination. a. Accepting to love a robot b. Accepting to be loved by a robot.

outcomes, where haptic, audio and motion played more important roles then smell, taste and color change.

4   Design

According to key design issues and questionnaire results, the function and form of the robot was designed. Love in itself is intangible and very personal. If we consider the Philia love, there is a sense of nurturing that is part of this relationship, and it is natural for us to be attached to someone or something that we nurture [7]. We wanted to evoke this nurturing capability in both humans and robots through our design and through this develop the sense of attachment that leads to Philia love.

4.1   Form and Function

Communicating any sense of feeling let alone a synthetic expression of love is quite a challenge, to say the least. To feel and express emotions, a combination of signals engaging all the senses is important. Sound, touch and vision would all play a part in the success of our intended solution. With such a monumental task set before us, we attempted to address all modalities in order to create a framework for an expression of synthetic love. With the hopes to develop a new emotion where being "human" is not necessarily a requirement, we started where we thought attraction would make its initial impact and developed a form that was pleasing not only to behold but to hold as well.

There are some key points that must be considered in designing the appearance of the robot. Firstly, the Lovotics robot should look timeless, representing no particular culture or generation. Children and adults should love and feel loved by it in equal measure. Secondly, there are differing tastes between genders. Males prefer objects with well defined straight lines or angles and in the case of love symbol, they prefer them to be less obvious or even better the lack of it. Females on the other hand can accept curvier, round and soft objects and obvious love cues being displayed [8], [9]. The appearance of the robot should be unisex using a simple and clean design. Males should love it more for its novel technology and females should love it more for its display of affection.

Referring to the results of our survey we found that middle-aged people believed that between human-to-human interpersonal affection, looks (physical appearance) and and touch (tactile feeling) were most important, closely followed by sound. In contrast, haptic and motion/position feedback was most important when discussing main sources for human-to-robot affection. The results of the questionnaire also showed that people preferred an organic shape to a geometric one for the robot. With these results taken into account, a form that was curvilinear and pleasing to the eye was developed. We made the robot quite compact, almost infantile or pet-like in size and dimension to instigate a feeling of smallness and fragility - something that one would like to nurture and take care of.

As it is supposed to fit in the hand, the size of the Lovotics robot should be small. Smaller things are likely perceived to be cute, less threatening and humans have the tendency to protect smaller things. Being small also allows the user to carry the robot with them. In addition to small, it should also be light in coherence with its size. A perception mismatch may be detrimental as all the positive feelings associated with being small may be diminished if the Lovotics robot is heavy. In addition, it is supposed to interact with the user at close proximity and it has to fit in the visual field of the user to capture his/her attention [10].

The choice of color can vary as different people like different colors. Pink or red could be a choice as they are representative of love [11],[12]. It is recommended that white or blue can be used as the primary color with some shades of other colors at some features. Blue is found to be the favorite color of many people and white is a color that induces trust [13]. In addition, the robot can be personalized by the user using accessories. By accessorizing the robot, the user can build up an ownership and develop more of a liking for the robot. This also allows the user to distinguish his/her robot from others.

4.2   Developed Robot

According to the proposed process, a robot was designed as presented in Fig. 3 and 4.

When a machine portrays itself as dependent, our nurturing instincts respond to this and this leads to an attachment with the machine [7]. The design of the form of the first Lovotics robot was inspired by the shape of an egg. The egg shape evokes our nurturing instincts, making us want to love and care for it. We used the feedback from the questionnaire to distill down to the basic characteristics that a love robot should embody and designed a robot to reflect these qualities. The tactile and audio senses as well as the appearance were the key elements. We have expressed these elements through the robot in the way it moves, changes shapes and produces sounds to express itself. We learned that it is a combination, sometimes subtle, of these elements that was understood by humans as the robot expressing its love.

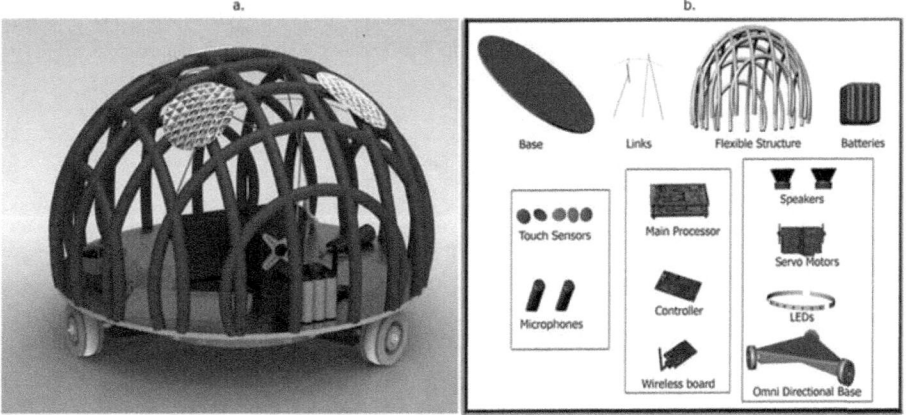

Fig. 3. a. Designed robot b. Hardware components

The robot is equipped with microphones and speakers for audio interaction with the environment. Also touch sensors are implemented on the surface of the robot for tactile interaction. The robot is connected to the server where all computation is processed.

The robot uses servo motors to actuate a vertical and tilting movement by virtue of applying force on its flexible endoskeleton design. Translational and Rotational motion is realized with a mobile base.

The presented robot is the first version of the Lovotics robot and initial user studies showed positive feedback from users regarding the developed robot. We are perusing our research of design and development of the robot and will examine our new design with proper user studies.

4.3 Behaviors

The robot is capable of six basic behaviors: Moving up and down to change heights, Rotation, Navigation in two dimensional space, vibration, changing color and tilting. Combination of these behaviors can be employed to express different emotions. These basic six behaviors are illustrated in Fig. 5.

5 Future Work

Our next step is to conduct an extensive qualitative fieldwork to help us gather more data about user responses to the present prototype we have developed. Using this information, we can further redesign the robot to better evoke the sense of nurture and care within the robot and human to further their love relationship. Along with the development of the robot itself, we look to understand and define the role of this new genre of robots in the social and cultural context. It is difficult to predict what the relationship will be between humans and robots

Fig. 4. Developed robot with its soft, spongy skin

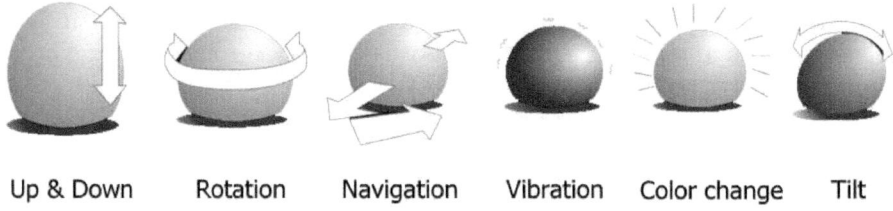

Up & Down   Rotation   Navigation   Vibration   Color change   Tilt

Fig. 5. Lovotics Basic behaviors

in the future, and tough questions such as 'how does this change our definition of love?' or 'is this kind of love appropriate?' need to be addressed. The way to do this is to continue this study in exploring 'love' and studying how man and machine are evolving into a new identity and relationship and to create a range of Lovotics robot to tackle these issues.

Acknowledgement

This research is carried out under CUTE Project No. WBS R-705-000-100-279 partially funded by a grant from the National Research Foundation (NRF) administered by the Media Development Authority (MDA) of Singapore.

References

1. Joachim, H., Rees, D.: Aristotle: The Nicomachean Ethics. Clarendon Press, Oxford (1951)
2. Lewis, C.: The four loves. Houghton Mifflin Harcourt, Boston (1991)

3. Soble, A.:Eros, agape, and philia: readings in the philosophy of Love. Paragon House Publishers (1989)
4. Norman, D.: Emotional design: Why we love (or hate) everyday things. Basic Civitas Books (2004)
5. Norman: Affordance, conventions, and design. Interactions 6(3), 38–43 (1999)
6. Mori, M.: The uncanny valley. Energy 7(4), 33–35 (1970)
7. Turkle, S.: A nascent robotics culture: New complicities for companionship, vol. 6, p. 2007 (2006) (On-line article, retrieved January)
8. Wellmann, K., Bruder, R., Oltersdorf, K.: Gender designs: aspects of gender as found in the design of perfume bottles. In: Design and emotion: the experience of everyday things, p. 87 (2004)
9. Xue, L., Yen, C.: Towards female preferences in design–A pilot study. International Journal of Design 1(3), 11–27 (2007)
10. Silvera, D., Josephs, R., Giesler, R.: Bigger is Better: The influence of physical size on aesthetic preference judgments. Journal of Behavioral Decision Making 15(3), 189–202 (2002)
11. McDonagh, D., Hekkert, P., van Erp, J., Gyi, D.: Design and emotion: The experience of everyday things. CRC, Boca Raton (2004)
12. Jirousek, C.: Psychological implications of color, http://char.txa.cornell.edu/language/element/color/color.htm
13. Hallock, J.: Colour assignment, http://www.joehallock.com/edu/COM498

Talking to Robots: On the Linguistic Construction of Personal Human-Robot Relations

Mark Coeckelbergh

Department of Philosophy, University of Twente
Postbox 217, 7500 AE Enschede, The Netherlands
m.coeckelbergh@utwente.nl

Abstract. How should we make sense of 'personal' human-robot relations, given that many people view robots as 'mere machines'? This paper proposes that we understand human-robot relations from a phenomenological view as social relations in which robots are constructed as quasi-others. It is argued that language mediates in this construction. Responding to research by Turkle and others, it is shown that our talking *to* robots (as opposed to talking *about* robots) reveals a shift from an impersonal third-person to a personal second-person perspective, which constitutes a different kind of human-robot relation. The paper makes suggestions for empirical research to further study this social-phenomenological process.

Keywords: human-robot relations, philosophy, phenomenology, language, construction, interpretation.

1 Introduction

The field of social robotics and human-robot relations is a growing and attractive inter-disciplinary research field [1][2][3][4]. Robots are not only more intelligent and autonomous; they are also more capable of interaction with humans. Some use the terms 'social robots' [5] or 'artificial companions' [6] and suggest a near-future scenario of widespread 'living with robots': robots will enter the personal sphere and provide companionship, entertainment, sex, and health care.

Whether or not this scenario will actually happen, there are already people who live with robotic 'companions' such as robot pets and there have been experiments with relations between robots and elderly people and between robots and children [7]. These experiences and observations raise many philosophical and scientific issues.

One striking observation is that people often address robots in a very 'personal' way. Their language use is similar to that in a human-human relation. We start talking *to* robots, not just *about* robots. For instance, Turkle and others report that one of the residents of a nursing home says to the robotic doll My Real Baby: "I love you. Do you love me?" [8]. How can we make sense of this kind of 'personal' and 'social' language use, given that many people view robots as 'machines'?

This paper reflects on the use of language in human-robot relations by developing the following argument.

M.H. Lamers and F.J. Verbeek (Eds.): HRPR 2010, LNICST 59, pp. 126–129, 2011.

2 Appearance, Language, and the Construction of Artificial Others

Approaching human-robot relations from a phenomenological point of view enables us to attend to the appearance of robots to human consciousness. We do not always perceive robots as 'mere objects' or 'machines'; robots can appear to us as 'quasi-others' [9] and human-robot relations as quasi-social relations. This is happens in particular with robots that are highly interactive and have a human-like (e.g. child-like) or animal-like appearance. Sometimes robots appear as more-than-things and this constitutes a particular kind of human-robot relation that is formally or structurally similar to a social human-human relation [10]. For instance, we play a particular role (e.g. the mother of the robot), develop social expectations (this are typically a kind of 'second-order' social expectations: e.g. I expect the robot to want something from me, it expects me to do something), we ascribe emotions to the robot ("it seems not very happy now"), and adapt our behaviour based on expectations (e.g. we expect the robot to be unhappy when we perform a particular action so we decide not to do it).

How can we best conceptualize the role of *language* in this social-phenomenological process? Our use of language mediates in at least two ways: (1) it does not merely represent but also *interprets* the robot and the human-robot relation and (2) it also helps to *construct* the robot as quasi-other and the human-robot relation as a social relation. This happens mainly by means of a shift from an impersonal third-person perspective ("it") to a personal second-person perspective ("you"). Instead of only thinking "it is happy now" we might address the robot and say "you are happy now, aren't you?" If this happens, a (stronger) social human-robot relation is being constructed. Sometimes the first-person plural is used ("we"), especially in cases of joint action.

In these cases, the robot is addressed and related to as if it were a human person. Here language functions not as a representation of a (social) 'objective' reality; it interprets whatever that reality is and helps to construct it. To say that there is 'first' the quasi-social relation and *then* language use adapted to this relation is not an adequate description; instead, language use is an integral *part* of the social relation and shapes it. For example, if I address a child robot (or, for that matter, a baby) as a 'you' instead of an 'it' this language use is part of the developing relation between me and the robot, a development that becomes more 'personal'. By talking to the robot in this way, the relation is constituted *as* a social relation. Thus, instead of seeing language use only as emerging from the human-robot relation as an 'objective' state of affairs (a representation of the relation), language gets a more 'active' interpretative and constructive role.

This linguistic-phenomenological framework helps us to make sense of existing research results, for example those offered by Turkle et al. Rather than interpreting the robots in question mainly as 'evocative objects' [11] that function as a stand-in for a human, the proposed framework reveals 'personal' robots as linguistically constructed artificial others. Here human identity is not delegated to the robot; instead, the robot and the human-robot relation are given their own, distinct and unique identity by means of narratives and pronouns that interpret and construct the robot as a quasi-other and constitute the relation as a social relation in a particular context.

How exactly the robot is addressed (talked to) and talked about will depend – among other things – on the appearance of the robot in a particular context and on the personal history of the human (for example the human might have never seen that type of robot and therefore experience uncertainty about how to address it). But language use plays an active role in defining 'what the robot is' and is not just a reflection or representation of the social relation.

3 Suggestions for Empirical Research

This claim can be turned into a hypothesis for empirical research: the language we use when talking about and *to* robots is not only a *result* of what happens in the human-robot relation but also constructs that relation. In order to test this claim, we could set up an experiment in which the *linguistic* 'environment' is manipulated in such a way that the human-relation is pre-defined and then observe what happens to the relations. For instance, in one series of interactions the instructor could pre-define the relation by using an impersonal third-person perspective when talking about the robot ("it") and in another series the instructor would encourage a second-person perspective by addressing the robot with "you". It is expected that there will be a significant difference in how people interact with and relate to the robot.

Note also that in order to study human-robot relations as *relations* one would need to shift the usual focus on (a small number of) *interactions* to long-term relational developments. This would reveal human-robot relations as changing and developing - as our interpretations and constructions of these relations also continuously change and develop.

4 Conclusion

If we wish to enhance our understanding of the 'personal' and 'social' dimension of what goes on between humans and robots, both philosophical and empirical work could benefit from more attention to the linguistic mediation of human-robot relations – in particular their interpretation and construction. The discussion offered in this paper needs further development but sketches a tentative conceptual framework that could guide further reflections and research in this direction.

References

1. Breazeal, C.: Toward sociable robots. Robot. Auton. Syst. 42, 167–175 (2003)
2. Dautenhahn, K., et al.: What is a robot companion – Friend, assistant, or butler? In: Intelligent robots and systems, IEEE/RSJ International Conference on In Intelligent Robots and Systems (2005)
3. Dautenhahn, K.: Methodology and themes of human-robot interaction: A growing research field. Int. J. Adv. Robot. Syst. 4(1), 103–108 (2007)
4. Levy, D.: Love and sex with robots: The evolution of human-robot relationships. Harper, New York (2007)
5. Breazeal, C.: Toward sociable robots. Robot. Auton. Syst. 42, 167–175 (2003)

6. Floridi, L.: Artificial intelligences's new frontier: Artificial companions and the Fourth Revolution. Metaphilosophy 39(4-5), 651–655 (2008)
7. Turkle, S., Taggart, W., Kidd, C.D., Dasté, O.: Relational artifacts with children and elders: The complexities of cybercompanionship. Connection Sci. 18(4), 347–361 (2006)
8. Turkle, S., et al.: Relational artifacts with children and elders: The complexities of cybercompanionship. Connection Sci. 18(4), 347–361, 357 (2006)
9. Ihde, D.: Technology and the lifeworld. Indiana University Press, Bloomington/ Minneapolis (1990)
10. Coeckelbergh, M.: Personal Robots, Appearance, and Human Good: A Methodological Reflection on Roboethics. Int. J. Soc. Robot. 1(3), 217–221 (2009)
11. Turkle, S., et al.: Relational artifacts with children and elders: The complexities of cybercompanionship. Connection Sci. 18(4), 347–361 (2006)

Using Empathy to Improve Human-Robot Relationships

André Pereira, Iolanda Leite, Samuel Mascarenhas, Carlos Martinho, and Ana Paiva

IST - Technical University of Lisbon and INESC-ID,
Av. Prof. Cavaco Silva, Taguspark 2744-016, Porto Salvo, Portugal
{andre.pereira,iolanda.leite,samuel.mascarenhas,
carlos.martinho,ana.paiva}@inesc-id.pt

Abstract. For robots to become our personal companions in the future, they need to know how to socially interact with us. One defining characteristic of human social behaviour is empathy. In this paper, we present a robot that acts as a social companion expressing different kinds of empathic behaviours through its facial expressions and utterances. The robot comments the moves of two subjects playing a chess game against each other, being empathic to one of them and neutral towards the other. The results of a pilot study suggest that users to whom the robot was empathic perceived the robot more as a friend.

Keywords: human-robot interaction, companionship, empathy.

1 Introduction

Robots are becoming part of our daily lives. The application domains where robots interact socially and cooperate with humans as partners, rather than as tools, is increasing. The more robots can socially interact with humans, the more people will be willing to accept them in public spaces, workplaces and even their homes. The LIREC Project (Living with Robots and Interactive Companions)[1] aims to create a new generation of interactive and emotionally intelligent companions (robots or embodied virtual agents) that are capable of establishing long-term relationships with humans.

If robots are to become our companions, then their social requirements must be addressed in order to make future robotic systems acceptable, usable and engaging. We argue that one of such social requirements is empathy, which involves perspective taking, the understanding of nonverbal cues, sensitivity to the other's affective state and communication of a feeling of caring [7]. In social psychology, the internal process of empathy is not clearly defined yet, and thus some definitions of empathy overlap with the concepts of emotional contagion (or mimicry), sympathy and pro-social behaviour [2].

Wispé [17] defines empathy as "an observer reacting emotionally because he perceives that another is experiencing or about to experience an emotion". But

---

[1] http://www.lirec.org/

M.H. Lamers and F.J. Verbeek (Eds.): HRPR 2010, LNICST 59, pp. 130–138, 2011.

some authors go even further, arguing that empathy not only includes affective processes, but also cognitive and pro-social behaviours (for example actions taken to reduce the object of distress) [2]. As such, empathy is often related to helping behaviour and friendship: people tend to feel more empathy for friends than for strangers [10].

Research on empathic agents is divided in two main branches: agents that simulate empathic behaviour towards the users and agents that foster empathic feelings on the users [12]. Previous research shows that agents expressing empathy are perceived as more caring, likeable, and trustworthy than agents without empathic capabilities, and that people feel more supported in their presence [4].

The main purpose of this paper is to investigate users' perceptions of a robotic companion with empathic behaviour, more specifically in terms of the possible relation of friendship established between them. To do so, we developed a scenario where a social robot watches, reacts empathetically and comments a chess match played by two humans. In this paper, we present the results of a pilot study that we conducted as a first step to evaluate this hypothesis.

2 Related Work

Similar to [3,9], our goal is to develop an artificial companion capable of establishing and maintaining a long-term relationship with users. Concerning this goal, the study presented in this paper is centered on how the display of empathic behavior affects the way humans perceive their social relationships with robots or artificial agents. In this section, some work on robots and virtual agents displaying empathic behavior will be presented.

Most work conducted with empathic robots only addresses one aspect of empathy, namely emotional contagion, where the user's affective state is mimicked. For instance, in [8], a study is conducted with an anthropomorphic robot that uses speech emotion recognition to decide the user's emotional state and then mirrors the inferred state using a corresponding facial expression. In another recent study [14], a robot with the form of a chimpanzee head, mimics the user's mouth and head movements.

Different from the aforementioned work, we do not propose to express empathy just by mimicking the user's facial expressions. Instead, we took inspiration from the field of virtual agents, where other forms of empathic behaviour were implemented. For instance in [13], an animated agent assists users in an application for job interview trainning, predicting the user's affective state through physiological signals. The user answers job-related questions while the agent says empathic statements of concern, encouragement or congratulation to users. These forms of empathic statements are also used in our work. However, we do not determine the user's affective state using physiological sensors. Instead, a role-taking approach to empathy is proposed, where the robot projects itself into the user's situational context to determine the user's affective state and the resulting empathic response. A similar approach was proposed in [15], where a model of empathy that involves self-projection was implemented, but only considering empathy between synthetic characters and not towards users.

3   Modelling Empathy

Empathy can be seen as a process mainly composed by two phases. The first phase includes the assessment of the other's affective state, and in the second phase the subject reacts taking into account the other's state (either by affective responses or more "cognitive" actions). Therefore, to model empathic capabilities in social robots we need to (1) recognize the user's affective state and (2) define a set of empathic behaviours to be displayed by the robot taking into account the user's state. The focus of this paper is on the second part of the empathic process.

In order to model empathic and non empathic behaviors in our robot, we have applied some of the characteristics referred in [6] as attitudes of empathic teachers that can induce empathy and understanding on students. Even though we do not intend to make our robot act like a teacher but as a game companion, our work was inspired by Cooper's comparison between empathic and non empathic teaching behaviors. This comparison was obtained by interviewing and observing teachers and students in the classroom. The behaviours are grouped by the following components: body-language, positioning, content of teaching, method of teaching, voice, attitudes, facial characteristics and responses. Given the limitations of our application scenario (robot's embodiment, technology, etc.), we only modelled characteristics from the last two components: facial characteristics and responses.

4   Case Study

To evaluate the influence of different empathic behaviours on user's perceptions of a robotic companion, we developed a scenario where Philip's iCat [16] observes the game of two humans playing chess, reacting emotionally and commenting their moves (see Figure 1). The iCat treats the two players differently: it exhibits empathic behaviours towards one of them - the *companion*, and behaves in a neutral way towards the other player - the *opponent*. These behaviours are reflected on the robot's facial expressions and utterances, as will be shown in the next subsections.

This scenario is a follow-up work of a previous scenario in which the iCat played chess against a human opponent [11]. To avoid the conflict between expressing empathy and acting as an opponent, in this scenario we placed the robot in an outside position. Also, having two players interacting at the same time allows us to simultaneously evaluate the two different conditions in the iCat's behaviour (empathic and neutral).

4.1   Interpreting the User's Affective State

Our previous work on affect recognition [5] highlighted the importance of contextual information to discriminate some of the user's states. In the particular context of a chess game, we identified a set of contextual features related to

Fig. 1. Two users playing chess with the iCat observing the game

the state of the game that are relevant to discriminate user's valence (positive or negative) and engagement to the robot. Therefore, to simulate an empathic process in our robot, its affective state will depend on the state of the game in the perspective of the companion (which ultimately is related to his/her affective state). We are aware that the iCat's affective states may not reflect accurately the affective state of its companion. However, when humans try to understand the affective states of each other, there are also many factors that blur this evaluation.

When a new move is played on the chessboard by one of the players, the iCat's affective state changes. The new board position is evaluated using a chess evaluation function in the perspective of the iCat's companion, which means that it will return positive scores if the companion is in advantage (higher values indicate more advantage), and negative scores if the companion is in disadvantage. Such evaluation values are the input of the *emotivector* system, an anticipatory mechanism which generates an affective state based on the mismatch between an "expected" and a "sensed" value. The emotivector system can generate nine different affective states, and each affective state is associated to a different facial expression in the iCat's embodiment. For more details on the emotivector system and its implementation in the iCat please consult [11]. The iCat's mood is also influenced by the state of the game, which is reflected in the robot's facial expressions in a similar way as it was done for our the previous scenario.

4.2 Empathic versus Neutral Behaviours

Inspired on the characteristics of empathic teachers cited before, we defined two sets of utterances for each affective state of the iCat: "empathic" utterances, to be used when the iCat is commenting the companion's moves, and "neutral" utterances, to be used when the robot is commenting on the opponent's moves. While neutral utterances merely indicate the quality of the move in a very direct way (e.g. "bad move", "you played well this time", ...), empathic utterances often contain references to possible companion's emotions, and try to encourage and motivate the companion (e.g. "you're doing great, carry on!").

As an example, suppose that the companion is loosing the game and plays a bad move; the consequent iCat's affective state is "expected punishment" (meaning that the current state is bad, as the robot was expecting). In this situation, a possible comment of the iCat would be "don't be sad, you didn't had better options". After that, if the opponent plays a good move and captures one of the companion's pieces, the iCat may say to the opponent "good move", even though its facial expressions and mood will reflect the negative affective state (empathic towards its companion). The iCat is also empathic to the companion by using his or her name two times more than it does when speaking to the opponent.

Two other empathic mechanisms were implemented. First, when players are thinking on the game, the iCat looks at the companion two times more than it looks at the opponent. Second, the iCat congratulates the companion when she/he captures a piece and encourages the companion in critical moments of the game, weather he/she is gaining advantage or disadvantage (for example, when the chances of winning become evident).

5   Experiment

The goal of the described experiment was to evaluate if users to whom the iCat behave more emphatically perceived the robot more as a "friend" than users to whom the iCat was neutral.

5.1   Procedure

The experiment was performed with undergraduate students from IST - Technical University of Lisbon. Ten participants between 22 and 24 years old, all of them male, played a total of five games. Subjects had never interacted with the iCat robot before and all of them knew how to play chess at a beginner level.

Two different conditions regarding the iCat's behaviour were evaluated as described earlier: *empathic* (for subjects playing with the black pieces) and *neutral* (for participants playing with the white pieces). At the beginning of the experiment, participants were asked to chose a side of the board and sat down. Before they started playing, some instructions were given regarding the experiment: they had to play an entire chess match, having the iCat next to the chessboard commenting their game. Participants were not informed about the differences in the iCat's behaviour. At the end of the experiment, they were asked to fill a questionnaire and were rewarded with a movie ticket.

5.2 Experimental Measures

For this experiment we wanted to measure the participant's perceived friendship towards the iCat robot. Mendelson [1] reviewed several existing friendship questionnaires and identified six relevant, conceptually distinguishable functions: (1) *stimulating companionship* - doing enjoyable or exciting things together; (2) *help* - providing guidance and other forms of aid; (3) *intimacy* - being sensitive to the other's needs and states and being open to honest expressions of thoughts, feelings and personal information; (4) *reliable alliance* - remaining available and loyal; (5) *self-validation* - reassuring, encouraging, and otherwise helping the other maintain a positive self- image; (6) *emotional security* - providing comfort and confidence in novel or threatening situations. From these descriptions and based in the context of our scenario, we defined two affirmations for each dimension (see Table 1). Participants expressed their agreement or disagreement about these affirmations using a 5 point Likert scale.

Table 1. Questions used in our friendship questionnaire

Dimension	Questions
Stimulating Companionship	I enjoyed playing chess with the iCat observing the game. I would like to repeat this experience.
Help	iCat helped me during the game. iCat's advices/comments were helpful for me.
Intimacy	iCat shared its affective state with me. iCat showed sensibility towards my affective state.
Reliable Alliance	I would trust iCat's opinion for guiding me in a future game. iCat was loyal to me.
Self-Validation	iCat encouraged me to play better during the game. I felt more confident playing with the iCat.
Emotional Security	iCat provided comfort in the difficult moments of the game. During difficult moments of the game, iCat's support was useful to me.

5.3 Results and Discussion

By comparing the friendship questionnaire in both conditions, we obtained some interesting results. For each dimension and for each participant we calculated the mean of the two items that composed that dimension. Figure 2 contains the average of each dimension from the participants of each condition.

With the exception of the *help* dimension, all other dimensions were rated higher in the empathic condition. This dimension is related to the helping behavior displayed by the iCat after every user's move. The addition of empathic reactions to this behavior does not seem to affect the helping behaviour of the companion.

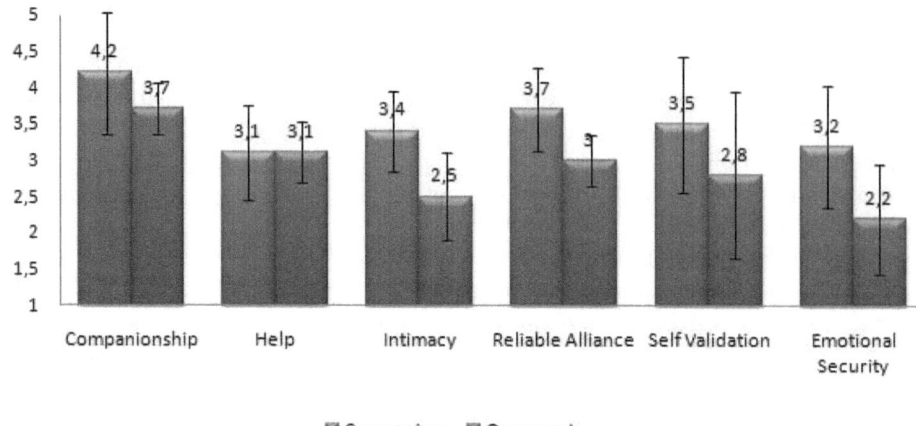

Fig. 2. Mean values of each friendship questionnaire dimension for the two conditions (error bars indicate the standard deviation)

Several dimensions had higher ratings in the empathic condition: participants agreed that the robot provided *emotional security* in the difficult moments of the game and claimed an increased sense of *intimacy* because of the shared robot's affective state. In both conditions, subjects considered the robot as a game *companion* as they both enjoyed playing with the iCat by their side. But even in this condition we could find a difference for better in the empathic condition.

6   Conclusions and Future Work

This paper addressed role of empathic behaviours in social robots that attempt to establish long-term relationships with humans. Our assumption is that if users perceive a robot as an empathic entity, they can more easily build some kind of friendship relation with them. The results of the preliminary experiment suggest that the participants with whom the iCat behaved in an empathic manner considered the robot friendlier. By looking separately at the friendship dimensions of the employed questionnaire, we retrieved more interesting findings. Intimacy, reliable alliance, self validation and emotional security dimensions had higher ratings in the empathic condition. The companionship dimension was also slightly higher in the empathic condition.

Modelling empathic behaviors in social robots seems to be relevant to improve the interaction with users. We intend to strengthen these results by performing a larger study with more participants to further determine the relevance of each friendship dimension on the user's perceived relationship with the robot, and which empathic behaviors have more influence on each dimension.

Acknowledgements. The research leading to these results has received funding from European Community's Seventh Framework Program (FP7/2007-2013) under grant agreement n 215554, and by 3 scholarships (SFRHBD/41358/2007, SFRH/BD/41585/2007, SFRHBD/62174/2009) granted by FCT.

References

1. Measuring friendship quality in late adolescents and young adults: Mcgill friendship questionnaires. Canadian Journal of Behavioural Science 31(1), 130–132 (1999)
2. Empathy: Its ultimate and proximate bases. The Behavioral and brain sciences 25(1), 1–72 (2002)
3. Bickmore, T., Picard, R.: Establishing and maintaining long-term human-computer relationships. ACM Transactions on Computer-Human Interaction (TOCHI) 12(2), 327 (2005)
4. Brave, S., Nass, C., Hutchinson, K.: Computers that care: investigating the effects of orientation of emotion exhibited by an embodied computer agent. Int. J. Hum.-Comput. Stud. 62(2), 161–178 (2005)
5. Castellano, G., Leite, I., Pereira, A., Martinho, C., Paiva, A., McOwan, P.: It's all in the game: Towards an affect sensitive and context aware game companion, pp. 1–8 (September 2009)
6. Cooper, B., Brna, P., Martins, A.: Effective affective in intelligent systems - building on evidence of empathy in teaching and learning. In: Paiva, A.C.R. (ed.) IWAI 1999. LNCS, vol. 1814, pp. 21–34. Springer, Heidelberg (2000)
7. Goldstein, A.P., Michaels, G.Y.: Empathy: development, training, and consequences / Arnold P. Goldstein, Gerald Y. Michaels. L. Erlbaum Associates, Hillsdale (1985)
8. Hegel, F., Spexard, T., Vogt, T., Horstmann, G., Wrede, B.: Playing a different imitation game: Interaction with an Empathic Android Robot. In: Proc. 2006 IEEE-RAS International Conference on Humanoid Robots (Humanoids 2006), pp. 56–61 (2006)
9. Kidd, C., Breazeal, C.: Robots at home: Understanding long-term human-robot interaction. In: IEEE/RSJ International Conference on Intelligent Robots and Systems, IROS 2008, pp. 3230–3235. IEEE, Los Alamitos (2008)
10. Krebs, D.L.: Altruism: An examination of the concept and a review of the literature. Psychological Bulletin 73(4), 258–302 (1970)
11. Leite, I., Martinho, C., Pereira, A., Paiva, A.:icat: an affective game buddy based on anticipatory mechanisms. In: Padgham, L., Parkes, D.C., Müller, J., Parsons, S. (eds.) AAMAS, vol. (3), pp. 1229–1232. IFAAMAS (2008)
12. Paiva, A., Dias, J., Sobral, D., Aylett, R., Sobreperez, P., Woods, S., Zoll, C., Hall, L.E.: Caring for agents and agents that care: Building empathic relations with synthetic agents. In: AAMAS, pp. 194–201. IEEE Computer Society, Los Alamitos (2004)
13. Prendinger, H., Ishizuka, M.: The Empathic Companion: a Character-Based Interface That Addresses Users' Affective States. Applied Artificial Intelligence 19(3-4), 267–285 (2005)
14. Riek, L., Robinson, P.: Real-time empathy: Facial mimicry on a robot. In: Workshop on Affective Interaction in Natural Environments (AFFINE) at the International ACM Conference on Multimodal Interfaces (ICMI 2008). ACM, New York (2008)

15. Rodrigues, S., Mascarenhas, S., Dias, J., Paiva, A.: I can feel it too!: Emergent empathic reactions between synthetic characters. In: Proceedings of the International Conference on Affective Computing & Intelligent Interaction, ACII (2009)
16. van Breemen, A., Yan, X., Meerbeek, B.: icat: An animated user-interface robot with personality. In: Proceedings of the Fourth International Joint Conference on Autonomous Agents and Multiagent Systems, AAMAS 2005, pp. 143–144. ACM, New York (2005)
17. Wispé, L.: History of the concept of empathy. Cambridge University Press, Cambridge (1987)

Author Index